T0255353

CAMBRIDGE BIOLOGICAL SERIES

GENERAL EDITOR :—ARTHUR E. SHIPLEY, M.A.
FELLOW AND TUTOR OF CHRIST'S COLLEGE, CAMBRIDGE

THE NATURAL HISTORY
OF
SOME COMMON ANIMALS

THE NATURAL HISTORY
OF
SOME COMMON ANIMALS

BY

OSWALD H. LATTER, M.A.

SENIOR SCIENCE MASTER AT CHARTERHOUSE, FORMERLY
TUTOR OF KEBLE COLLEGE, OXFORD

CAMBRIDGE:
at the University Press
1904

CAMBRIDGE UNIVERSITY PRESS
Cambridge, New York, Melbourne, Madrid, Cape Town,
Singapore, São Paulo, Delhi, Mexico City

Cambridge University Press
The Edinburgh Building, Cambridge CB2 8RU, UK

Published in the United States of America by Cambridge University Press, New York

www.cambridge.org
Information on this title: www.cambridge.org/9781107619524

First published 1904
First paperback edition 2013

A catalogue record for this publication is available from the British Library

ISBN 978-1-107-61952-4 Paperback

PREFACE

THE main object of this volume is to set forth the natural history of those animals which usually serve as types of animal structure in elementary courses of zoology. The smaller of these have been omitted, and only those whose habits and gross structure can be observed without the assistance of a microscope have been discussed. The Dragonfly and Wasp, as a rule, find no place in the courses to which allusion has been made. They are here introduced, partly to bring forward the phenomena of metamorphosis, which is not to be found in the Cockroach, and partly owing to the interesting nature of their life-histories.

Sixteen years' experience of teaching zoology has convinced me that we have been too closely wedded to structure and have wrongly divorced function from our elementary courses of instruction. Structure alone is very liable to become dry bones in very deed, and consequently to fail to attract that interest without which good work is almost impossible. It has been my endeavour throughout this book to present each animal to the reader as a *living* thing, a machine of whose wonderful parts and workings we have obtained some little knowledge, but concerning which there is yet much to be ascertained. In the hope that a desire for further knowledge may be kindled in the minds of some of those into whose hands the book may fall, appendices have been added containing lists and brief diagnoses of the non-marine relatives of the animals dealt with in the several chapters. These lists have been confined to British species, but are not of equal range in all cases, for reasons which will be apparent to those acquainted with the groups. While a personal knowledge of anatomy will, no doubt, enable the reader more fully to appreciate the facts that are mentioned, I have attempted to make the accounts of the various animals intelligible and interesting to those who have not the advantage of technical knowledge.

In addition to those to whom acknowledgments of kind assistance are made in the text, I must express my thanks to Mr M. D. Hill, of Eton College, for reading the proofs and for numerous helpful suggestions, to my former pupil Mr Edgar Schuster, of New College, Oxford, for valuable help in consulting the literature of the subject, and to Mr A. E. Shipley, of Christ's College, Cambridge, for permission to use a large number of the plates that appear in Shipley and MacBride's *Zoology*, and for frequent advice and help in the course of the preparation of the book. I am also much indebted to Mr Edwin Wilson for the skill and care which he has bestowed upon my original drawings, figs. 15, 16, 20, 34, 35, 36, 37 and 38, in preparing them for reproduction.

O. H. L.

CHARTERHOUSE,
GODALMING.
April 1904.

CONTENTS

CHAPTER		PAGES
I.	Earthworms and Leeches	1—32
	British Oligochæte Worms . . .	33—47
II.	The Crayfish	48—70
III.	The Cockroach	71—92
	British Orthoptera	92—98
IV.	Dragonflies	99—112
	British Dragonflies	112—122
V.	Wasps	123—159
	British Wasps	159—161
VI.	The Fresh-water Mussel	162—202
	British Fresh-water Lamellibranchs .	202—204
VII.	Snails and Slugs	205—234
	British Land and Fresh-water Gastropods	235—245
VIII.	Frogs, Toads, and Newts	246—292
	British Amphibia	292—301
IX.	Some Common Internal Parasites of Domestic Animals	302—321
	Index	323—331

LIST OF ILLUSTRATIONS

FIG.		PAGE
1	Earthworm, *Lumbricus terrestris*	4
2	Setæ of " "	5
3	Section of " "	8
4	Internal organs of *Lumbricus terrestris*	16
5	Leech, *Hirudo medicinalis*	27
6	Internal organs of *Hirudo medicinalis*	28
7	Internal organs and limbs of Crayfish	50
	Plan of gastric ossicles of Crayfish	54
8, 9, 10, 11, 12	Diagrams of Kreidl's experiments on statocysts of *Palæmon*	61—64
13	Cockroach, *Periplaneta orientalis*, dorsal view	73
14	" " " side view	73
15, 16	Front leg and antenna comb of Wasp, *Vespa germanica*	77
17	Mouth-appendages of *Periplaneta*	81
18	The internal organs of "	83
19	The metamorphosis of a Dragonfly	108, 109
20	Wings of Dragonflies	113
21, 22	Hind wing of Wasp	126
23	Stridulating organ of Mosquito	133
24	Sting of Wasp, *Vespa germanica*	135
25	Section of sting, " "	136
26	Tip of sting-needle of *Vespa germanica*	138
27	Nest of Wasp at an early stage	142
28	" " a late "	147
29	Right valve of *Anodonta cygnea*	165

FIG.		PAGE
30	Dorsal view of *Anodonta mutabilis*	168
31	Right side of „ „	169
32	„ „ „ „ dissected	175
33	Sections of „ „	180
34	Glochidium of „ „ posterior view	189
35	„ „ „ ventral „	190
36	Views of young Mussel	194
37	Views of shell of *Unio*	194
38	Young Mussel seven weeks old	195
39	*Helix pomatia*, animal and shell	206
40	Section of head of *Helix*	218
41	Radula of *Helix*	218
42	Pulmonary chamber of *Helix*	221
43	Internal organs of *Helix*	227
44	Hyoid apparatus of Frog	266
45	Skeleton of Frog	268
46	Nervous system of Frog	272
47	Urino-genital organs of Frog, male	278
48	Manus of Frog, male and female	280
49	Tadpoles of Frog	285
50	*Molge cristata*, the Warty Eft	296
51	Views of skull of Newt	299
52	Liver-fluke, *Distomum hepaticum*	304
53	Tapeworm, *Tænia solium*	312
54	*Trichina spiralis*	317

CHAPTER I.

EARTHWORMS AND LEECHES.

EARTHWORMS occur commonly in nearly all parts of England. Their body is roughly cylindrical, but tapers towards each extremity. It is divided into a number of similar rings or segments, about 140—180, by constricting furrows. The surface is moist and iridescent; this latter property is an optical effect produced by the fine striations with which the delicate cuticle is engraved, and is not due to the presence of any pigments. Earthworms inhabit burrows in the surface of the earth, and for the most part limit their operations to the top 12 or 18 inches (*i.e.* in the soil which is richest in decomposing vegetable and animal substances and in which decomposition occurs most rapidly), but during periods of prolonged drought or frost they descend to greater depths and undergo æstivation or hibernation, as the case may be, coiled up into a compact spiral and lying in a small excavated chamber. This is lined with small stones which prevent close contact with the surrounding earth and so permit free respiration.

L.

The sides of the burrow are kept moist by slime discharged from the glandular cells of the skin, and perhaps by liquid discharged from the body-cavity through the dorsal pores which occur in the grooves that separate segment from segment. The slime is said to possess antiseptic properties, and thus to preserve the skin of the worm from harmful bacteria.

The mouth of the burrow is guarded by small stones or more frequently by one or more leaves pulled in to a greater or less distance. Fir-needles, stalks of horse-chestnut leaves and other similar things are often to be seen standing nearly erect upon the ground, their lower ends having been forcibly dragged into the mouth of a burrow by a worm. On still, warm nights in early autumn the rustling noise of fallen leaves being dragged along by worms is often plainly audible in favourable localities. Darwin has pointed out[1] that worms exhibit considerable intelligence in drawing the narrow end of leaves of various shapes foremost into the burrow: leaves with broad bases and narrow apices are generally pulled in tip first, whereas when the base is narrower than the apex the reverse position is usually found. There is no doubt that worms can judge which end of any leaf is the better to seize. The reason for thus pulling objects into the entrance of the burrow is probably to prevent the entry of foes, centipedes, parasitic flies, etc., to keep the burrow moist by preventing evaporation, to keep out the cold lower strata of air at night, to bring food supplies within safe reach, and also to enable the worm to lie near the mouth

[1] *Vegetable Mould and Earthworms.* London, 1881.

of the burrow unobserved. Here however they are not secure from all attack, for the quick ears of the thrush and other birds enable them to detect the slightest movement and, with a quick plunge of the beak, to seize and, after a brief tug-of-war, to extract the worm from its refuge. Frequently the well-known worm-castings are thrown up on the surface, and when this is so, leaves are not, as a rule, drawn into the burrows, the heap of castings serving the purpose.

The burrow is made partly by the awl-like tapering anterior end pushing aside the earth on all sides, and partly by the actual swallowing of the earth as the worm advances, so that the animal literally eats its way into the soil. The organic material in the swallowed soil serves as food, and the residue in a state of very fine division passes out at the anus, and is used either to form the above-mentioned castings or as a lining to the burrow, especially where this passes through hard, coarse earth.

Perfectly healthy worms seldom leave their burrows completely except perhaps after very heavy rain. The majority of those so frequently found travelling over the surface of roads and paths after rain are infected by the larvæ of parasitic flies and doomed to die. On warm, moist evenings, however, worms may be seen in hundreds lying stretched on the surface of the ground with only the broad flattened posterior end remaining in the burrow. Here we see one of the uses of this modification in the shape of the hinder segments of the body: their greater width enables them to obtain a firm purchase on both sides of the burrow, and thus the worm is provided with a sure anchor on which it can pull and at the slightest alarm shoot back like stretched elastic into the security of its burrow. At other times the flat tail is employed trowel-wise in smoothing the excrement against the walls of the burrow or in disposing the castings on this side and on that of the mouth of the burrow.

Locomotion is primarily effected by the alternate rhythmic contractions of the longitudinal and of the circular muscles of the body-wall which contract and elongate successive regions of the cylindrical body. The eight bristles (setæ or chætæ) with which each segment is furnished are however of scarcely less importance, inasmuch as they serve as so many little cogs to catch in irregularities of the surface and thus bring about movement in a definite direction. Muscles are attached to the inner parts of the setæ and can cause them to shift their positions; an arrangement that

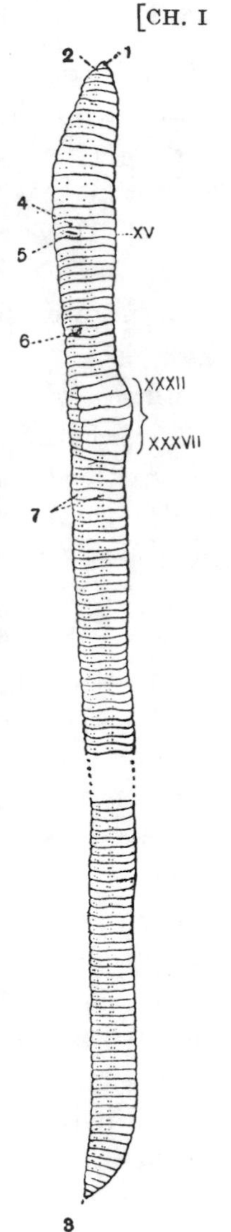

Fig. 1. Latero-ventral view of *Lumbricus terrestris*, slightly smaller than life-size. From Hatschek and Cori.

1. Prostomium. 2. Mouth. 3. Anus. 4. Opening of oviduct. 5. Opening of vas deferens. 6. Genital chætæ. 7. Lateral and ventral pairs of chætæ.

xv, xxxii and xxxvii are the 15th, 32nd and 37th segments. The 32nd to the 37th form the clitellum.

Fig. 1.

is very necessary in the rapid retrograde movement of a worm darting back into its burrow. If a worm is placed on a highly polished horizontal surface the contractions of its body do not result in any definite movement but merely in writhings. On the other hand, worms can climb along over nearly vertical surfaces provided these are rough. I have seen them moving at a good speed up a "dry wall" built of Bargate stone—a coarse-grained sandstone whose surface is admirably adapted to afford

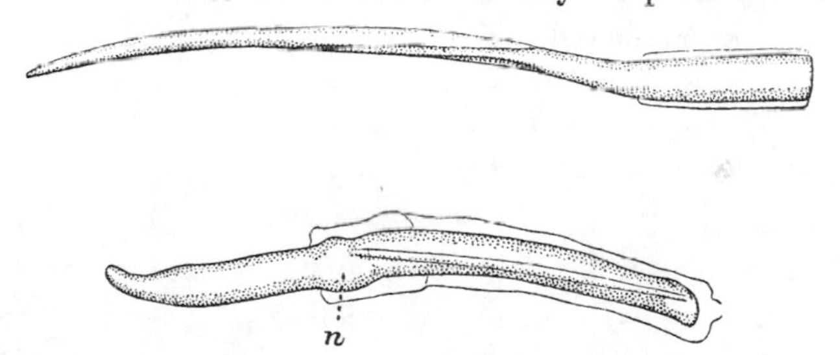

Fig. 2. A genital seta (grooved) from 26th segment of *L. terrestris*; actual length 1·75 mm.; and, below, an ordinary seta of the same individual; actual length 1·42 mm. The faintly indicated tube round the right-hand (inner) portion of the figures represents the seta sac.

hold for the numerous setæ. If a living worm be held in the hand the manner of using the setæ in ordinary locomotion is easily seen: as the body is elongated the setæ are retracted and disappear from sight, but directly the contraction of the longitudinal muscles of the body-wall begins the setæ shoot forth from their pits and can be seen plainly by the unaided eye and their points felt by the skin of the hand.

Digestion. In addition to the food obtained from the soil worms devour leaves, both fresh and decaying, and also animal substances, fat being especially attractive to them. By the sucking action of the muscular pharynx and the lips leaves are drawn to the burrow, and over the surface is discharged an alkaline fluid which softens and discolours the leaf-tissue, and appears to effect the partial digestion of both protoplasmic contents and starch granules. The fluid is derived from glandular cells in the skin of the anterior part of the body. Fragments of the softened leaf are sucked off by the worm, there being no teeth, and swallowed.

In passing along the œsophagus the food encounters the secretion from the calciferous glands. In *Lumbricus terrestris* there are two pairs of these glands in addition to a pair of œsophageal pouches; in *Helodrilus* (*Allolobophora*) *longus* the latter alone are present. It is probable that the primary function of these glands is to excrete calcareous matter derived from leaves. It is well known that such matter accumulates in the leaf-tissue and is not withdrawn with other substances when the leaf is detached from the parent tree, but is, as it were, excreted by the plant in the falling leaf. The majority of the leaves devoured by worms are such as have fallen in the natural order of things, and moreover all vegetable mould is very largely the product of decayed leaves. Hence worms are likely to take in a large amount of calcareous matter, and since they form neither shell nor bone there is no outlet for this substance and some special excretory apparatus seems necessary. At the same time it is probable that the calcareous fluid discharged into the œsophagus does play a part in digestion in serving to neutralise more or less the organic acids resulting from decay of vegetable substances in the soil, and thus assists the digestive action of the alkaline fluid which is poured over the leaves, and which continues its action subsequent to the act of swallowing. Nevertheless the contents of the gizzard and intestine are as a rule acid, probably as a result of fermentations occurring in the later stages of digestion. Harrington[1] finds that excess of acid in the food increases the amount of lime produced by the glands.

The crop serves as a temporary storage place in which the digestive processes continue prior to the passage of the food into the gizzard. Here the strong muscular walls and horny lining, aided by small particles of stone that are almost invariably to be found in this cavity, grind and triturate the food so that the contents of the intestine are always in a state of extremely fine division. Miss Greenwood[2] has described retractile cilia upon the epithelium of the intestine, and states that digestion is effected by the secretion of unicellular glands over the whole of the typhlosole and corresponding part of the intestine. The typhlosole, which forms so conspicuous a fold along the dorsal wall of the anterior portion of the intestine, is a contrivance for increasing the area of the internal absorbing surface without increasing the external bulk of the tube. The cylindrical body of a worm does not permit of enlarging the absorptive surface by augmenting the length of the intestine and throwing it into loops—an arrangement found in the convoluted intestines of most vertebrate

[1] *Journn. Morph.* vol. xv. suppl. 1900.
[2] *Journn. Physiol.* vol. XIII.

animals. A somewhat similar typhlosole is found in the intestine of fresh-water Mussels, though here it is accompanied by a few convolutions.

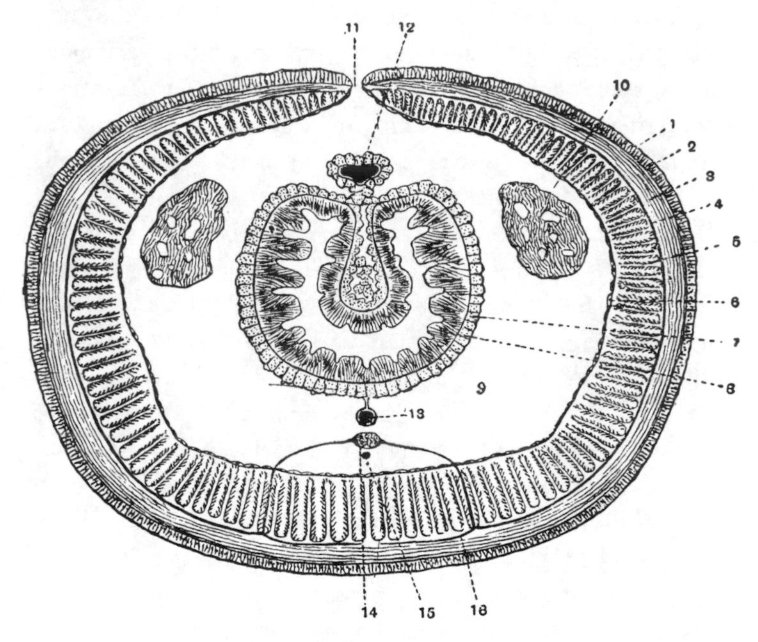

Fig. 3. Transverse section through *Lumbricus terrestris* in the region of the intestine, and of a dorsal pore. Magnified.

1. Cuticle. 2. Ectoderm or epidermis. 3. Circular muscles. 4. Dorsal nerve. 5. Longitudinal muscles. 6. Somatic epithelium. 7. Splanchnic epithelium or yellow cells. 8. Epithelium lining the intestine. 9. Cœlom. 10. Nephridium cut in section. 11. Dorsal pore. 12. Dorsal blood vessel lying along the typhlosole or the groove in the wall of the intestine. 13. Subintestinal blood vessel. 14. Ventral nerve cord. 15. Subneural blood vessel. 16. Ventral nerve.

The dorsal and ventral nerves are added diagrammatically. The other structures are drawn from nature.

The spiral valve that occurs in the intestine of the

Dogfish, and other Elasmobranch fishes, and the somewhat similar structure in the intestine of the medicinal leech and again the radial "mesenteries" of Sea-anemones and their allies are mechanisms adapted to securing the same end as the typhlosole, viz. that a tube whose internal surface shall greatly exceed its external in superficial extent.

Respiration and Circulation. In the absence of special organs of respiration this function is in the worm performed by the outer surface of the body. In each segment of the body a pair of blood vessels is given off from the subintestinal longitudinal trunk to the body-wall and skin, where the interchange of oxygen and carbon dioxide is effected through the moist surface of the integument. The oxygenated blood is then returned either to the subneural trunk or to the vessels which run in the walls of the anterior portions of the digestive system, the blood being kept in motion by the pairs of contractile "hearts" in segments 6 or 7—11. These hearts run from the longitudinal dorsal vessel round the œsophagus to the subintestinal vessel. Astonishment is sometimes expressed that worms can breathe underground. But the soil is seldom so closely packed that there is not a fair supply of air entangled in the spaces between the particles of earth, and doubtless the aëration of the soil is substantially enhanced by the very burrowing of the worms themselves. At the same time it is noteworthy that their blood is provided with a special vehicle of oxygen —hæmoglobin—not indeed contained in corpuscles as in the blood of vertebrate animals, but dissolved in the liquid itself; and, speaking generally, it is the case that such oxygen-carriers are found either in animals whose bulk is considerable in proportion to their respiratory surfaces, or in animals which though small live in an environment poor in oxygen, *e.g. Tubifex* living in mud, and the aquatic larvæ of some species of *Chironomus* (Gnat). It is very probable that the worms so often seen on the surface of the ground after heavy rain have come up to avoid the suffocation to which they assuredly are exposed when the soil becomes saturated and much of the included air expelled by the water. Many worms, however, normally live in very moist places and can withstand prolonged immersion in water. In order to test their powers of endurance in the absence of oxygen I placed one worm in a flask full of water that had been boiled to expel all gases and then allowed to cool in an atmosphere of carbon dioxide: in 3 minutes the worm was asphyxiated and to all appearance dead. A second worm was placed in ordinary tap-water that had not been boiled and though it made strenuous efforts to get out it was still alive and vigorous after 1½ hours' immersion; on transferring it to the other flask it too in a couple of minutes lay motionless. I then poured off the water and filled the flask with oxygen gas: the second worm recovered in ten minutes, and the first, which had been drowned for nearly 1¾ hours, in about double that time.

Excretion. The chief excretory organs of the body are the nephridia. These are fine, much convoluted tubes, opening internally by ciliated funnel-shaped mouths into the body-cavity and externally on the ventral surface.

Each segment of the body, with the exception of the first three and last one, possesses a pair of these organs. The funnel-shaped mouth lies in the segment anterior to that which contains the rest of the tube, the first part of the duct perforating the intervening septum. Each tube is divisible into several regions; the portion next to the funnel is ciliated, the next part glandular, and the third section, which functions as an expellent bladder, muscular. The tubes are abundantly supplied with blood, from which they eliminate waste substances in the form of uric acid. In this work they are assisted by the "yellow" or chloragogen cells which form so conspicuous an external covering to the intestine, especially along the line of the typhlosole, the groove of which structure is closely filled with them. The apparent connexion of the chloragogen cells with the alimentary system is accidental. They are in reality closely applied to the outer walls of the dorsal blood vessel and its intestinal offsets. They originate from the amœboid corpuscles of the cœlomic fluid which occupies the body-cavity (cœlom). Certain of these attach themselves to the walls of the blood vessels and extract from the blood yellowish-brown substances and thereby become converted into chloragogen cells. When filled with these coloured bodies they detach themselves from the walls of the blood vessels and float about in the cœlomic fluid. Their contents then break down into the blackish *débris* which is often found in the nephridial tubes and is by them conducted to the outside (Kükenthal[1], and Claparéde[2]). Foreign substances such as indigo,

[1] *Jena. Zeit.* vol. XVIII. 1885. [2] *Zeit. wiss. Zool.* XIX. 1879.

carmine, or iron injected into the body are also excreted by the activity of the chloragogen cells (Kowalevsky[1], Schneider[2]). In worms which have been starved and in which, presumably, there is a deficiency of waste products these cells are pale and dull in colour.

The amœboid corpuscles of the cœlomic fluid have a remarkable power of attacking bacteria and other microscopic organisms such as Gregarines and Infusorians or even small Nematode worms. If such parasites enter the cœlom the amœboid cells surround and destroy them. Their operations are however not confined to the inside of the earthworm. The slime of the body surface is in part composed of mucus secreted by the skin, and in part of cœlomic fluid and its corpuscles which find exit through the dorsal pores. The corpuscles are thus able to attack and destroy bacteria before they effect an entry into the body. There is no doubt that a worm is constantly exposed to these minute organisms for the upper layers of the soil teem with them. The slime itself is a protection, for it both arrests the bacteria and holds them stranded in the trail which the worm leaves behind it in its progress. The application of a grain of some irritant, such as corrosive sublimate, enables one to see how a worm protects itself. Immediately the irritant touches the skin the segments in front and behind the seat of injury are forcibly constricted while the affected segment itself swells up in consequence of the increased pressure brought to bear upon it from both sides. At the same time there is

[1] *Biol. Centralbl.* IX. 1890. [2] *Zeit. wiss. Zool.* LXI. 1895–6.

a conspicuous gush of cœlomic fluid from the dorsal pores in that region and an abundant secretion of mucus from the skin itself. Thus the threatened region is, as it were, isolated by ligatures from the rest of the body and all the defensive resources at once brought to bear upon the enemy. The cœlomic fluid is alkaline and contains crystals of calcium carbonate, and also contains micro-organisms which when isolated and reared in artificial cultures emit the characteristic smell of earthworms. It is therefore not improbable that this odour is due to the micro-organisms and not really a feature of the worm itself (Lim Boon Keng)[1].

Special senses. Worms have no organs of special sense in the usual acceptation of these words. They are absolutely deaf to sound though very sensitive of vibrations of the soil, or of the surface on which they are resting, which fact at times causes it to appear that they do hear certain notes, whereas in reality the note sounded has but caused the surface in question to vibrate, and thus has affected the worm. They possess no eyes, and yet they can appreciate light and darkness and quickly move out of a powerful light into shade. This sensitiveness to light appears to reside only in the anterior regions of the body. In some experiments which I conducted by flashing with a small mirror a spot of light, obtained by an electric lamp, on to a worm lying in the shade, a slight sensitiveness seemed to coincide pretty closely with the darker pigmentation that is so frequently observable on the first 15 or 20 segments of large worms, but the prostomium and first

[1] *Phil. Trans.* CLXXXVI. B. pt. 1.

segment exhibited a very marked appreciation of the light, being far superior to the other anterior segments in this respect. If the spot of light was thrown a few inches ahead of the moving worm, a sudden start took place directly the prostomium entered the small illuminated area, and the animal withdrew or turned aside: whereas if the light was directed on to any of the first dozen segments, the worm merely paused. Hesse[1], however, finds that the entire surface is sensitive to light, perception being especially acute at the two extremities. Langdon[2] has described sense-cells as occurring freely all over the skin, in great numbers at the front and hind ends, but it by no means follows that these are sensitive to light, they may be merely tactile in function. The sense of smell is undoubtedly possessed by worms, for they will find scraps of favourite food when hidden out of sight, and assemble in quantities in the soil beneath lumps of fat thrown out on the ground. They may often be observed to raise the front end of the body and wave it in this or that direction in the air, as though they were obtaining information that reached them through the air, *i.e.* as odours.

The prostomial lobe, or upper lip, seems to act as a special tactile, or perhaps also taste organ, and may often be seen thrust forward in a tentative manner and speedily withdrawn by telescoping into the segments behind if the object explored be distasteful. No doubt the whole surface of the body is highly sensitive to touch, and to certain

[1] *Zeit. wiss. Zool.* LXI. 1895–6. [2] *Journ. Morph.* XI. 1895.

influences which we perhaps should designate as tastes, *e.g.* a grain of salt touching any part of the worm's body causes severe discomfort and strong pungent smells as of ammonia are perceived equally well by all parts. Such stimuli as these can, however, not be regarded as natural, and it is not inappropriate to recall the fact that while we ourselves *taste* salt and *smell* ammonia we are keenly aware of their presence on any moist thin portion of our skin, *e.g.* the surface of the eyeball, or on spots whence the skin has been abraded.

Reproduction. The worm is hermaphrodite, possessing both male and female organs, and when sexually mature may be known by the development of the 'cingulum' or 'clitellum,' the thickened ring girdling the body for some 6 or 7 segments and popularly regarded as the scar marking recovery from bisection with a spade! There is no very clearly defined breeding season, for worms may be found in the act of mating at any time from spring to autumn, provided the weather be suitable; warm, moist days being the most favourable. In *L. terrestris* the reproductive organs consist of two pairs of spermathecæ in segments 9 and 10 respectively, two pairs of testes in segments 10 and 11 contained in a median vesicula seminalis, and three pairs of lateral vesiculæ seminales in segments 9, 10 and 11. In the vesiculæ the spermatozoa mature, and from them are conducted out of the body by the vasa deferentia opening on the 15th segment. There is a pair of ovaries in segment 13; the ova are passed out of the body by the short oviducts, whose internal apertures are in segment 13, the external in segment 14. The ova rest for a longer or shorter period in small sacs situated upon the sides of the oviducts. In the mating of such species as pair above ground, two worms from adjacent burrows, each retaining a firm hold in its own burrow by means of its flattened tail, apply their ventral surfaces to one another so as to overlap for about a third of the length of the body. The head of each worm points toward the tail of the other. The clitellum of each secretes a band of mucus which binds the two worms firmly together, so firmly, indeed, as to cause two well-marked constrictions, while a slimy covering, the slime-tube, surrounds the two worms from the 8th to the 33rd segments. The seminal fluid, containing spermatozoa and spermatophores, flows within the slime-tube, and, during sexual union, in the early stages of the formation of the cocoons spermatophores cover the dorsal and lateral surfaces of segments 9, 10 and 11 of each worm and are packed between the two

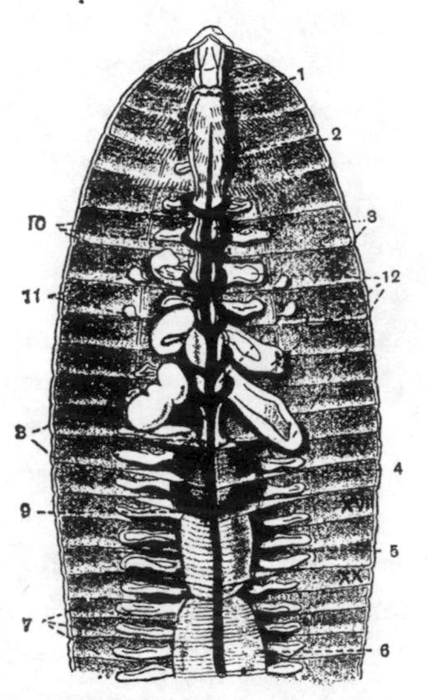

Fig. 4. Anterior view of the internal organs of an Earthworm, *Lumbricus terrestris*. Slightly magnified. From Hatschek and Cori.

1. Central ganglion or brain. 2. Muscular pharynx. 3. Œsophagus. 4. Crop. 5. Muscular gizzard. 6. Intestine. 7. Nephridia (the reference lines do not quite reach the nephridia). 8. Septa. 9. Dorsal blood vessel. 10. Hearts. 11. Spermathecæ. 12. Vesiculæ seminales.

The Roman figures refer to the number of the segments.

worms. The spermatozoa flow backwards from the male aperture in a longitudinal groove on each side to the receptacula (spermathecæ) of the other worm, the grooves of the two animals together forming a temporary tube. Hence only one worm can emit spermatozoa at any given time, otherwise there would be opposing currents. The worms are so placed (*Lumbricus terrestris*) that the 9th segment of each is opposite the 32nd (1st clitellar) of its mate, then the thickened clitellum forms a barrier, past which no flow of seminal fluid can take place. In this position the modified and grooved genital setæ of the 26th segment come opposite the small aperture in the 15th segment, and are probably used to hold the lips of this opening apart during sexual congress. The long genital setæ in the "tubercula pubertatis" of the clitellum, and of segments 10 to 15, are probably used, the former to liberate the cocoon from its seat of origin, and the latter series to hold the cocoon off the ventral surface in the region of the oviducal openings and those of the spermathecæ, and thus allow ova and spermatophores to pass into the cocoon as it passes forwards. These specialised setæ replace those of ordinary form as the worm reaches maturity. The eggs do not pass out of the oviduct till near the end of the act of mating. Each of the two worms forms a cocoon, and slips out of it backwards, passing it forward over its head. The cocoon being elastic closes its two open ends as soon as the body of the worm is withdrawn, and becomes more or less lemon-shaped, its bulging centre being occupied by about 4 eggs, spermatozoa and albuminous material produced by the so-called capsulogenous glands, which may be seen on the ventral side of some of the segments in front of the clitellum. The cocoons, at first white but soon becoming yellow, are left in the earth, and as a rule only one of the contained eggs produces a young worm. The size of the cocoons differs in the various species, those of *L. terrestris* are from 6 to 8 mm. long by 4 to 6 mm. broad, of *Eisenia fœtida* from 4 to 6 mm. long by 2 to 3 mm. broad. There is some doubt as to the precise function of the spermathecæ. It seems certain that the spermatozoa contained in them are derived from some other worm. It is also the case that these organs are full of spermatozoa prior to sexual union, and are empty subsequent to that act, at any rate when cocoons are formed and eggs deposited. Worms have been observed to separate without producing cocoons, and though perhaps in some instances the separation may have been due to disturbance caused by observation, yet there is reason to think that two unions are necessary, one to fill the spermathecæ, and a second to form cocoons. In such a case it is probable that each worm acts as a carrier of spermatozoa from its first to its second mate, *i.e.* worm A gets its spermathecæ filled by the spermatozoa of B in the first union, and passes these spermatozoa to C in the second. The actions are probably often reciprocal. According to Goehlich[1] while spermatozoa are flowing from one worm to the spermathecæ of

[1] Semper's *Zool. Beitr.* II.; cf. also Wilson, *Journ. Morph.* IV.; K. Foot, *Journ. Morph.* XIV. 1898, and *Zool. Bull.* II.; Hering, *Zeit. wiss. Zool.* VIII.; Bergh, *Zeit. wiss. Zool.* XLIV and L. 1890; Rosa, *Bull. Mus. Zool. Anat. Torino*, IV.; Vejdovsky, *Entwick. Untersuch. Prag.* 1888—1892; Britscher, *Biol. Centralbl.* XXI. 1901.

the other, there is given out from the spermathecæ of the former a small quantity of mucus which hardens when it reaches the air: a second portion of mucus containing a group of spermatozoa is then emitted, this becomes attached to the first mass, and with it forms a spermatophore. The whole spermatophore is attached to the body of the other worm close to the clitellum. When the cocoon is made the spermatophores are rubbed off into it as the animal withdraws itself.

Light could probably be thrown on this matter by some such experiments as follow: keep a number of worms, each in a separate flower-pot, from infancy to maturity; kill[1] a few and examine the contents of their spermathecæ (it is conceivable that a worm may be able to pass spermatozoa into its own spermathecæ); allow remainder to mate once, note if cocoons are deposited; kill some and examine contents of spermathecæ; allow rest to mate a second time, pairing some with their former mates and others with different mates: kill all and examine spermathecæ.

Recuperative powers. A worm if cut in two can under favourable conditions recover, each portion ultimately forming a complete worm. It is also possible to cause the two severed portions to reunite, or to graft the head portion of one worm on to the tail portion of another. It is doubtful if under natural conditions such unions of severed parts ever occur, but it is, at any rate, clear that a worm can recover from the effects of wounds such

[1] The best way to kill worms is to plunge them for a moment in boiling water.

as must frequently be inflicted by birds and others of its natural foes.

Friedländer[1] has shown that not only may a number of anterior or posterior segments be regenerated, but also the supra-œsophageal ganglia, and other parts of the nervous system. The regenerating tissue consists of masses of amœboid cells, into which new growths make their way. The new parts are always of reduced diameter and exhibit frequent abnormalities. According to Hescheler[2] in *L. terrestris, L. rubellus, H.* (*A.*) *longus, E. fœtida, H.* (*A.*) *caliginosus* complete regeneration of the anterior end is only accomplished when comparatively few segments are lost: if the number exceed ten, some four or five segments only are formed anew. Monstrous individuals of *L. terrestris* and of *E. fœtida* have been described by Bell[3], Marsh[4], Williamson[5] and others. The monstrosity in all cases consists in a doubling of the hinder region so that the tail appears forked. The abnormality affects all the chief internal organs. The branching usually takes place at about three-quarters of the length from the head. Each branch ends in a functional anus.

Enemies and Parasites. Birds are proverbially the worst enemies of earthworms, and chief among them may be mentioned the thrush and rook, which are both equal to devouring a full-grown worm, while robins, hedge-

[1] *Zeit. wiss. Zool.* LX. 1895.

[2] *Vierteljahrschr. Nat. Gesellsch. Zurich.* XLII. 1897.

[3] *Ann. Mag. Nat. Hist.* (5) XVI.

[4] *Amer. Nat.* XXIV.

[5] *Ann. Mag. Nat. Hist.* (6) XIII.; cf. also Broom, *Trans. N. H. Soc. Glasgow*, II. *N. S.*

sparrows and many other small birds will greedily pick up young worms when the chance occurs. Sea-gulls too at certain seasons will follow the plough and eat such worms as they may find exposed. It is unnecessary to enumerate all the species of worm-eating birds; at one period of the year or another a large number of our native birds have recourse to a diet of worms.

Below the surface too the worm is not free from attack, being pursued by the mole, one of the most voracious feeders known; by the small shell-bearing slug *Testacella* (occasionally too by the large black slug *Arion ater*), by the swift, flat-bodied, red-brown centipede *Lithobius*, and by the larva of the beetle *Steropus madidus*, which will "bolt" worms from their burrows as a ferret does rabbits.

In addition to the mole, other mammals such as shrews and hedgehogs may be regarded as fairly constantly preying upon the worm, while many cold-blooded Vertebrates, such as toads, frogs, and lizards devour them readily.

Among insects the chief foes are beetles of the genus *Ocypus*, commonly known as "Devil's Coach-horses," whose normal food is for the most part composed of worms. Ants may occasionally be seen attacking a worm, but probably they have in such cases availed themselves of the opportunity afforded by a worm already reduced to comparative helplessness by other circumstances.

Aquatic worms fall frequent victims, of course, to numerous species of fish, leeches, and turbellarian flatworms. Certain dipterous flies deposit their eggs in the body of the living worm, and the resulting larvæ sooner or later destroy their host.

In the body-cavity, and in or on the septa, there may

constantly be found quantities of parasitic Nematodes (or thread-worms). These are the Rhabdites larvæ of *Pelodera pellio*, they do not become mature until they pass into the earth after the death of their host. Probably the young enter the worm through the mouth with food, but the life-history is not known with certainty.

R. Leuckart has described a species of *Ascaris* which lives free within the muscle fibres of the mole. The parasite has a well-marked boring tooth. Leuckart mentions that earthworms are infested by an *Ascaris* of similar appearance, and with a similar tooth. When eaten by a mole the young Ascarids of the earthworm are said to be destroyed by digestion. But when infected mole's flesh was eaten by a buzzard, the lungs and liver of the bird became covered with tubercles containing the young Ascarids, in the same stage of development as in the muscles of the mole. It is obvious that the life-history of this parasite requires re-investigation. Dujardin in 1838 and 1839 found in the "testicules" (*vesiculæ seminales*) of earthworms taken in Paris a Nematode which he named *Discelis filaria*; he described it as being white, thread-like, from 3 to 5 mm. long, blunt at the extremities; skin with faint transverse striations; tail with two disc-shaped suckers laterally. Dujardin subsequently failed to find this parasite among earthworms elsewhere, nor have they been since reported by other workers. In the ventral blood vessel of the earthworm occur, as first described by Cori, larval Nematodes, which von Linstow recognised as the immature forms of *Spiroptera turdi*, found as parasites in the coats of the stomach of the thrush, blackbird, redwing and some other birds, in which situations sexual maturity is reached. The birds no doubt get infected by devouring the earthworms: how these in their turn get the parasite is not known, but it is most probably by eating with the soil some of the disjecta of the bird[1].

Within the male genital organs, imbedded in the sperm mother-cells, is generally to be found the parasitic Gregarine Protozoon *Monocystis*, whose spore-laden cysts so frequently occur in hundreds in the cavities of the seminal vesicles. The spores (chlamydospores) probably pass with the spermatozoa into the cocoon, but how the fresh generation gains access to the worm is not known with certainty. It is noteworthy, as bearing alike on the life-history of this parasite and the reproductive processes in the worm itself, that the chlamydospores are found in the spermathecæ.

Economics. The effects produced on the surface soil by the action of earthworms have been most fully pointed out by Charles Darwin in his well-known book *Vegetable Mould and Earthworms.* It will be sufficient here to call attention to a few facts only. Worms play a most important part in maintaining the soil in a state suitable to vegetation. The burrows form ventilating tubes whereby the soil is aerated and respiration by the roots of plants rendered possible; at the same time they open up drainage channels, preventing the surface from becoming waterlogged. Doubtless also roots find an easy passage through the soil along the lines of burrows even after the walls

[1] Shipley, *Arch. d. Parasit.* VI. No. 4, 1902, where other references are given.

have more or less fallen in. Moreover, the excrementitious earth with which the burrows are lined is peculiarly suited to the root fibres, being moist, loose and fertile. Microscopic examination of the earth deposited by worms shows it to resemble two-year-old leaf-mould such as gardeners use for seed-pans and pricking-out young seedlings: most of the plant-cells are destroyed, shreds and fragments alone remaining, discoloured and friable, mingled with sand grains and brown organic particles. In chemical composition too worm-castings are very similar to fertile *humus*.

The castings which are thrown up on the surface materially improve the quality of the upper soil, and render it the more fit for the germination of seeds, many of which directly or indirectly get covered by the upturned earth. It has been reckoned that there are upwards of 50,000 worms in an acre of soil of average quality: hence the total effect of the work of this vast host must be very considerable. Each worm ejects annually about 20 ozs. of earth. The weights of earth thrown up in a single year on two separate square yards observed by Darwin were respectively 6·75 lbs. and 8·387 lbs., amounts which represent respectively 14·58 tons and 18·12 tons per acre per annum.

In addition to this tilling action worms improve the quality of the soil by the leaves and other organic *débris* which they drag into their burrows, and thus bring within reach of bacteria. These, as it is well-known, especially abound in the upper soil, and effect the speedy decomposition of dead animals and vegetable tissues.

Archæologists are indebted to worms for the preservation of many ancient objects, such as coins, implements, ornaments, and even the floors and remains of ancient buildings that have become buried by the soil thrown up as worm-castings. The process of disappearance is of course hastened by the excavations effected by the worms below the surface, for the collapse of the burrows slowly but surely allows objects on the surface to sink downwards.

In the disintegration of rocks, and the denudation of the land, worms play an important part. The penetration of the burrows, and the lining with castings, carries down the humus-acids to a considerable depth and exposes the underlying rocks to their solvent action. Within the body of the worm itself small stones and grains of sand are reduced to yet finer dimensions and rendered the more easy of transport by wind or water. On sloping surfaces the upturned castings, at first semi-fluid, flow down, and when dry roll down the incline, or are washed by the rain into the valleys and ultimately carried out to sea, while on level ground the dried castings are blown away to lower spots by the wind. The more or less parallel ridges that are frequently found on the sloping sides of grass-clad hills are in part, at any rate, formed by the material derived from worm-castings, which has temporarily lodged against tufts of grass, etc., and in turn furnished a richer and deeper soil for stronger growth which arrests yet more and so increases the ledge. All land surfaces, whether level or sloping, provided they are occupied by worms, are reduced in altitude by their action. In no small degree then may earthworms be held responsible for our valleys and hills and all the softer features of our scenery.

Leeches.

Leeches are flattened worms that may be regarded as more or less degenerate relatives of earthworms. The body is segmented, and each segment is further divided externally into smaller annuli. In the *Hirudinidæ* there are five annuli per segment except at the extremities. All our British species live in fresh water and are very reluctant to leave it, but in India and Ceylon and elsewhere leeches are found living on land among vegetation and capable of travelling at a fair rate in pursuit of animal blood.

In the United States leeches are bred for surgical purposes. At Newton, Long Island, there is, or was, a leech farm of some 13 acres extent. The farm consists of oblong ponds of about 1½ acres each and 3 feet or so deep. The bottom is covered with clay and the banks made of peat. The eggs are deposited in the peat from June onwards through the warm weather. The adults are fed once every six months on fresh blood placed in linen bags suspended in the water. The most destructive enemies are rats, which dig the cocoons out of the peat.

Locomotion. When in the water a leech either progresses by holding on to some surface by the anterior and posterior suckers alternately and looping its body after the fashion of a geometer caterpillar; or casting itself free from all attachment, both elongating and flattening its

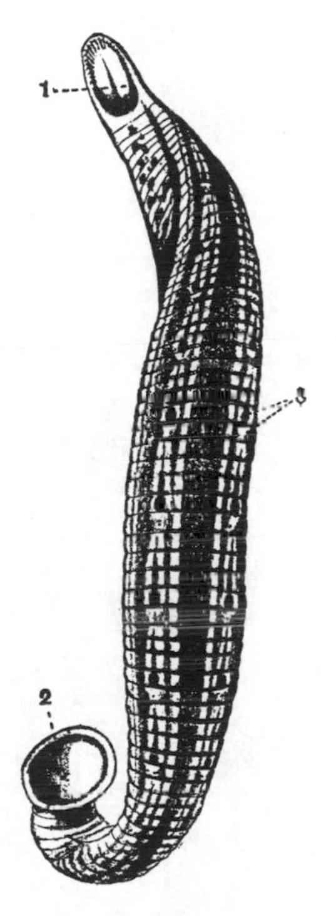

Fig. 5. *Hirudo medicinalis*, about life size.
1. Mouth. 2. Posterior sucker. 3. Sensory papillæ on the anterior annulus of each segment. The remaining four annuli which make up each true segment are indicated by the markings on the dorsal surface.

body to a band-like shape, swims by rapid and graceful undulations of its tough, muscular body.

Food. Blood of vertebrate or invertebrate animals constitutes the chief food: cattle, birds, frogs and tadpoles, snails, insects, small soft-bodied crustacea, and worms are all attacked by various species as occasion offers. The mouth may be provided with teeth as in the Medicinal Leech, which possesses three jaws resembling portions of a circular saw and studded on the circumference with numerous minute teeth, or there may be no teeth, but a protrusible proboscis-like pharynx.

It has been shown that certain blood-sucking leeches discharge over and into the wound inflicted by the teeth a fluid which has the property of preventing the blood from clotting, so that the flow of blood is not checked while the animal is feeding, and probably too solidification is prevented in the capacious sacculated crop into which the blood passes. Opportunities for a full meal occurring but intermittently, it is all important that the leech should be able to avail itself to the utmost of the rare occasion and lay up a store that will last for a considerable time. A well-fed Medicinal Leech, now but rarely found in this country, is said to be able to fast for some nine months. In this species the stomach is very small, and the intestine though straight contains a closely coiled spiral valve down which the food but slowly winds its way to the anus. The common Horseleech (*Aulostoma*) does not as a matter of fact suck the blood of vertebrates, having but

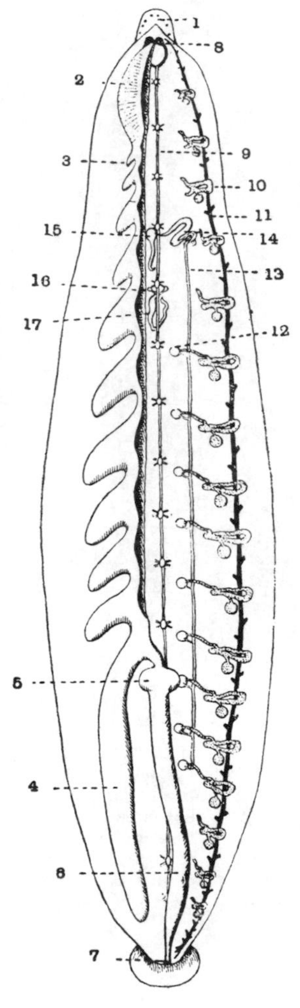

Fig. 6.

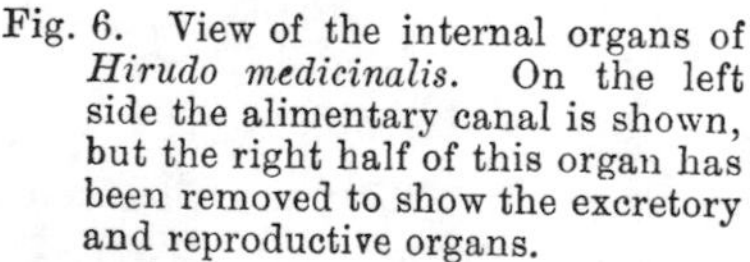

Fig. 6. View of the internal organs of *Hirudo medicinalis*. On the left side the alimentary canal is shown, but the right half of this organ has been removed to show the excretory and reproductive organs.

1. Head with eye spots. 2. Muscular pharynx. 3. 1st diverticulum of the crop. 4. 11th diverticulum of the crop. 5. Stomach. 6. Rectum. 7. Anus 8. Cerebral ganglia. 9. Ventral nerve-cord. 10. Nephridium. 11. Lateral blood vessel. 12. Testis. 13. Vas deferens. 14. Prostate. 15. Penis. 16. Ovary. 17. Uterus—a dilatation formed by the conjoined oviducts.

few and weak denticules. It subsists on worms and other soft-bodied aquatic animals.

The blood plasma is red in *Hirudo* and *Aulostoma* and *Nephelis*, colourless in *Clepsine* and its relatives. It contains numerous colourless amœboid corpuscles. There is a pair of large lateral longitudinal blood vessels in the former group, and in the latter a dorsal and a ventral vessel in addition. These all have muscular walls and are contractile. Besides these well-defined blood vessels there are also numerous irregular plexuses, capillary sinuses, traversing the tissues.

Respiration is effected not by any specialised organs but through the general surface of the skin, which is richly provided with capillary plexuses, whose branches extend between the inner ends of the cells of the epidermis. The waving motion that leeches frequently keep up when at rest is probably for the purpose of moving the surrounding water and so promoting respiration.

Nervous system and special senses. The nervous system consists of a pair of supra-œsophageal ganglia united to a paired ventral chain with ganglia in each segment. At the anterior and posterior ends concentrations of ganglia have occurred. From this central nervous system nerves pass to all parts of the body. There are scattered over the surface of the body, especially in its anterior region, small cup-shaped bodies to which nerves are supplied and which probably serve as organs of touch and perhaps also of smell—a sense which leeches undoubtedly possess. Round the edge of the anterior sucker there are in the Medicinal Leech ten minute black dots which are modifications of the above-mentioned tactile bodies and are sensitive to light. There is no evidence to show that there is any sense of hearing.

Excretion. Excretion is performed by means of nephridial tubes of which there are, in the Medicinal Leech, seventeen pairs, lying in the second to the eighteenth segments inclusive. The inner end of the tube is a spongy mass of ciliated cells lying in a space which probably represents a remnant of the reduced body-cavity: the external opening is on the ventral side of the body. Urea and uric acid have been found in the nephridia. It is probable that the granular pigmented cells of the "botryoidal" tissue, which occurs in abundance immediately within the muscular walls of the body, is also excretory in function and comparable to the chloragogen cells of the earthworm.

Reproduction. Leeches are hermaphrodite and many details of the processes of reproduction are yet to be discovered. In some species the eggs are deposited in cocoons formed by clitellar glands situated in the skin in the region of the genital apertures: in others the young are cared for by the parent and carried about attached to its body for a time. Observations[1] on *Hirudo geometra* (Linn.), a species parasitic on gudgeon, roach, and other fish, kept in aquaria showed that mating frequently followed the addition of fresh water to the vessels. The eggs were deposited on the glass sides of the aquarium about 24 hours after sexual union; they are oval in shape and reddish-brown in colour and covered by a white web-

[1] Brightwell, *Ann. Mag. Nat Hist.* 1842.

like cocoon. The eggs hatched in 30 days producing young $\frac{1}{3}$ inch long, which speedily fastened on to tadpoles. The adults died a few days after laying the eggs.

The Medicinal Leech and Horseleech deposit their eggs in capsules which are attached to submerged plants and similarly situated bodies, or are buried in the mud or in holes in the banks. The cocoon of the latter species[1] is oval, about half an inch long, of a greenish-brown colour. It consists of an outer, loose, spongy case surrounding the central more compact envelope of the inner chamber, within which is a dense semi-transparent gelatinous substance in which are embedded some ten pear-shaped eggs.

The cocoons are formed, as in earthworms, in the shape of rings open at each end and completely surrounding the segments by which they are secreted: the leech withdraws the anterior part of the body, sliding backwards out of the cocoon as out of a collar. It is said that the cocoon is fastened to some foreign object by means of the mouth.

A small but common leech, *Nephelis vulgaris*[2], which occurs in most ponds and sluggish streams, deposits brown capsules of eggs on the underside of the leaves of water plants among the eggs of fresh-water snails. The eggs are laid in June and hatch in about six weeks. This leech sucks the blood of snails, but small Crustaceans and even Rotifers are found within its digestive cavity. The habits of *Nephelis tesselata*[2] are somewhat different, the young to the number of 200 or more remaining attached to the

[1] Bowerbank, *Ann. Mag. Nat. Hist.* 1845.
[2] Brightwell, *loc. cit.*

parent by their posterior suckers for some time. Another common leech, *Clepsine complanata*[1], which preys on water snails such as *Planorbis*, nurtures the young in a similar way: here the eggs pass from the genital aperture in two longitudinal rows back to the hinder portion of the body which is looped up into a pouch to receive them and to constitute a brood chamber until the young hatch and fasten themselves to the ventral surface of the parent by their suckers.

Dispersal. In addition to the locomotive powers of the adult leeches whereby they can travel both in water and for short distances over land, currents and floods doubtless play an important part in carrying the eggs to fresh places. These modes of dispersal can hardly account for the wide distribution of many species; this has probably been effected by the aid of the animals on which the leeches prey. They have been seen clinging to the legs and other parts of cattle and horses that have entered the water to drink, and also to the legs and feet of water birds and waders, by whose agency they might well be carried to ponds and streams distant several miles in a few minutes. Moreover the mud adhering to the feet of such birds and mammals might at times contain in it eggs whose development would be in no way impaired by transport to other waters.

[1] Brightwell, *loc. cit.*

BRITISH OLIGOCHÆTE WORMS.

[The nomenclature, classification, and descriptions (in so far as necessary for British Oligochætes) here adopted are those given by Dr Wilhelm Michaelsen in *Das Tierreich*, 10 (Friedlander and Son, 1900).]

Family ÆOLOSOMATIDÆ. Septa incomplete or absent. Brain not separated from hypodermis. Asexual reproduction frequent.

genus *Æolosoma* Ehrbg. Skin usually with oil glands. Prostomium ventral, ciliated, not separated by a furrow from next segment. Setæ in four groups, two dorsal and two ventral. Testis in 5th, ovary in 6th, unpaired. Spermathecæ simple, 1 to 3 pairs in segments 3rd to 5th.

A. quaternarium Ehrbg. Oil glands orange red. Prostomium not broader than following segment. Nephridia absent from region of œsophagus. Total segments 7 to 10 : very small. In slime covering stones in fresh water.

A. beddardi Mchlsn. Oil glands absent. Prostomium broader than following segment. Length about 2·5 mm. Total segments 8.

A. hemprichi Ehrbg. Oil glands orange to carmine. Prostomium broader than following segment. Length 2 to 5 mm. Total segments 4 to 13. In wells, tanks, etc.

A. headleyi Beddard. Oil glands light green or bluish. Prostomium broader than following segment. Length about 2·5 mm. In fresh-water aquaria.

A. variegatum Vejd. Oil glands partly colourless, partly yellow or greenish-yellow. Prostomium broader than following segments. Nephridia absent from œsophageal region. Has occurred in Ireland.

A. tenebrarum Vejd. Dirty white. Oil glands faintly yellow. Prostomium broader than following segment. Nephridia present in œsophageal region. Length 5 to 10 mm. In deep wells.

Family NAIDIDÆ. Septa developed. Brain separated from hypodermis. Asexual reproduction frequent. Setæ in two or four groups per segment. Ventral setæ prong-forked and hooked: dorsal various. Testes in segments 5th or 7th, rarely in 8th and 9th: ovaries in 6th or 8th, rarely in 10th. Spermatheca in 5th or 7th.

genus *Paranais* Czern. Two dorsal and two ventral groups of setæ per segment. Segment 3 not longer than the rest. Dorsal and ventral setæ prong-forked and hooked. Testes in 8th and 9th, ovaries in 10th.

P. litoralis (Müll.) Örst. First dorsal setæ on segment 5: longer and more slender than ventral setæ. Margins of fresh and brackish water.

genus *Chætogaster* K. Baer. Two groups, ventral only, of setæ per segment. Setæ absent from segments 3 to 5. 3rd segment much lengthened. Testes in 5th. Ovaries in 6th. Fresh water.

C. crystallinus Vejd. Prostomium inconspicuous. Œsophagus nearly as long as pharynx. Transparent. Length 2 to 3 mm.

C. diaphanus (Gruith). Prostomium inconspicuous. Œsophagus short. Transparent. Total segments 14 or 15. Length 10 to 15 mm.

genus *Nais* Müll., em. Vejd. First dorsal setæ on segment 6: needle-shaped or hooked setæ as well as filiform setæ in dorsal groups. Testes in 5th. Ovaries in 6th. Fresh water.

N. heterochæta Benham. Brownish. 2 eyes present: dorsal groups of setæ with prong-fork setæ. Total segments 31 to 41. Length about 6·5 mm.

N. josinæ. Vejd. Reddish. No eyes. Forking of dorsal setæ of posterior segments inconspicuous. Length 6 to 8 mm. Scotland?

genus *Dero* Ok. Posterior end of body modified into a respiratory funnel with branchiæ. First dorsal setæ on 6th segment, rarely on 5th. Testes in 5th. Ovaries in 6th. Fresh water.

D. latissima Bousf. Respiratory funnel broader than long, subrectangular, devoid of dorsal lip. Branchiæ ribbon-like. Total segments 20 to 30.

D. perrieri Bousf. Respiratory funnel trefoil-shaped, devoid of dorsal lip. Branchiæ 2 pairs, cylindrical, three or four times longer than broad, posterior pair the longer. Total segments 25 to 35. Length 12 mm.

D. obtusa Udek. Respiratory funnel with dorsal lip, no secondary branchiæ or palps. Branchiæ 2 pairs, leaf-like, short. Total segments 45 to 50.

D. mülleri Bousf. Respiratory funnel with dorsal lip, no secondary branchiæ or palps. Branchiæ 2 pairs, subrectangular, broader than long. Total segments 70 to 95. Length about 13 mm.

D. limosa Leidy. Respiratory funnel with dorsal lip flanked by a pair of secondary branchiæ. Branchiæ 2 pairs, about half as broad as long. Total segments 48 to 60. Length 6 mm.

D. furcata Ok., Bousf. First dorsal setæ in 5th segment, prong-forked. Respiratory funnel large, with dorsal lip fused with lateral walls and carrying a pair of long secondary branchiæ. A pair of long slender palps on edge of funnel. Branchiæ 2 pairs, long, subcylindrical. Total segments 35.

genus *Bohemilla* Vejd. Filiform setæ with a row of fine spines on concave side. Ventral setæ prong-forked and hooked ; first dorsal on 5th segment, some very long and spined, others short and simple.

B. comata Vejd. Colourless, transparent. Generally with 2 eyes. Total segments 38. Length 4 to 6 mm. In swamps, streams, etc.

genus *Ripistes* Duj. Very long dorsal filiform setæ on 6th, 7th, and 8th segments : first on 6th. Fresh water.

R. macrochæta (Bourne). Long setæ from two to five in each group, rest of dorsal setæ very short. Lives in tubes or free swimming.

genus *Slavina* Vejd. Very long dorsal filiform, setæ on 6th segment only.

S. appendiculata (Udek). Yellowish. Skin covered with foreign bodies and numerous papillæ. Generally with 2 eyes. Total segments 35 to 60. Length 10 to 20 mm. In ponds.

genus *Stylaria* Lm. Dorsal setæ exclusively filiform, first on segment 6. Prostomium produced, tentaculiform.

S. lacustris (L.). Dorsal setæ 1 long and 1 or 2 short per segment. Eyes present generally. Total segments 25. Length 10 to 15 mm.

genus *Pristina* Ehrbg. Dorsal setæ exclusively filiform, first on 2nd segment. Prostomium produced. No eyes. Testes in 7th : ovaries in 8th (in P. leidyi : position in British sp. not known). Fresh water.

P. æquiseta Bourne. Filiform setæ of 3rd segment not lengthened. Total segments 18 to 21. Length 7 to 8 mm. Botanic Gardens, probably introduced.

P. longiseta Ehrbg. Filiform setæ of 3rd segment enormously prolonged. Total segments 17 to 20. Length 8 mm.

Family TUBIFICIDÆ. Prostomium rounded. Setæ in four groups, two dorsal, two ventral. Ventral setæ hooked, simple or two-pronged : dorsal setæ various. One pair of internal seminal funnels in segment in front of male aperture. Male apertures paired, in 11th segment generally, 12th sometimes. One pair of spermathecal apertures, a segment in front or a segment behind male aperture ; seldom absent. One pair of testes in segment in front of male aperture. One pair of ovaries in same segment as male aperture.

genus *Branchiura* Beddard, em. Mchlsn. Spermathecal in segment 10, male apertures in segment 11. Dorsal setæ filiform as well as hooked (simple or forked) or fan-like. Male aperture in 11th segment. No penis. Spermatophores not formed. Fresh water.

B. coccinea (Vejd). Bright red. No branchiæ. In clean, running water.

B. sowerbyi Beddard. Branchiæ on last fifty to eighty segments. Length 38 to 50 mm. In mud of Victoria-regia tank ; introduced.

genus *Vermiculus* Goodrich. Spermathecal in segment 10, male aperture in segment 11, both median, unpaired. Female apertures paired, in furrow between 11th and 12th segments. Testes in 10th, ovaries in 11th. Dorsal and ventral setæ all prong-forked and hooked.

V. pilosus Goodrich. Blood red. Length 25 to 38 mm. Seashore (Weymouth). Clitellum from 10th to 13th and part of 14th segment.

genus *Clitellio* Sav. Spermathecal in 10th, male apertures in 11th, paired. Penis without chitinous bars. Spermatophores in spermathecæ. Dorsal and ventral setæ all prong-forked and hooked.

C. arenarius (Müll.). Pale red. Clitellum from 10th to 12th segments. Total segments 64 to 120. Length 30 to 65 mm. Seashore.

genus *Limnodrilus* Clap. Spermathecal in 10th, male apertures in 11th. Penis with chitinous bars. Spermatophores in spermathecæ. Dorsal and ventral setæ all prong-forked and hooked.

L. hoffmeisteri Clap. Bright-red to brown-red. Length of penis bars about eleven times the breadth at proximal end: slightly curved. Total segments about 95. Length 20 to 50 mm. In rivers and streams.

L. udekamianus Clap. Pink anteriorly, yellow posteriorly, with brownish peritoneal spots in each segment. Penis bars straight, slightly widened distally, length about four times breadth at proximal end. Length 30 to 60 mm. In clear and in muddy waters.

genus *Tubifex* Lan., em. Mchlsn. Some of ventral setæ prong-forked and hooked; dorsal with two or more pointed hooks or imperfectly fan-like, and some filiform usually. Spermathecal in 10th, male apertures in 11th. Ovaries in 11th segment, female aperture in 12th. Spermatophores in spermatheca.

T. tubifex (Müll.) (usually known as *T. rivulorum*). Red. Length 30 to 40 mm. In mud of streams and ponds. Penis soft without chitinous bars. Very common. The worms project the hinder end of the body from their mud tubes into the water and maintain a waving motion to and fro for respiratory purposes. Often so abundant as to form red patches on the mud.

T. benedeni (Udek.). Red-grey to blackish-red. Penis with chitinous bars. Skin with numerous papillæ arranged in irregular rings. Length 35 to 55 mm. Seashore and below tide marks.

T. costatus (Clap.). Pale red. Penis with chitinous bars. Skin smooth, without papillæ. Dorsal setæ of 5th to 13th segments (generally) spatulate. Length 16 mm. Under stones and rubbish on seashore (Sheerness).

Family LUMBRICULIDÆ. One pair of internal seminal funnels in same segment as male apertures and generally a second pair in preceding segment. Setæ S-shaped, simple or prong-forked, 8 per segment, in 4 pairs, 2 ventral and 2 lateral. Male apertures 1 pair, generally in 10th, sometimes in 8th or 11th segment. Female apertures 1 or 2 pairs, generally in furrow between 10th and 11th. Spermatheca 1 to 5 pairs. Ovaries 1 pair in segment next behind male aperture and sometimes a 2nd pair in the following segment.

genus *Lumbriculus* Grube. Male apertures in 8th. Penis retractile. Female 2 pairs, in furrows between 9th and 10th, and 10th and 11th. Spermathecæ 4 pairs in 10th to 13th, simple. Ovaries 2 pairs, in 9th and 10th.

L. variegatus (Müll.). Red to dark brown, greenish anteriorly. Prostomium conical, curved, with pore at apex. Total segments 140 to 200 or more. Length 40 to 80 mm., thickness 1 to 1·5 mm. In mud and sand or among weeds in ditches and swamps.

genus *Stylodrilus* Clap. Male apertures on non-retractile penis in 10th, posterior to ventral setæ. Spermathecal apertures in 9th, 2 pairs of internal seminal funnels in 9th and 10th, 1 pair of ovaries in 11th.

S. vejdovskyi Benham. Orange-red. Segments 2 to 4 annulate. Setæ prong-forked, upper shorter than lower prong. Length about 25 mm. On roots of water plants. Cherwell, Thames, etc.

Family ENCHYTRÆIDÆ. Spermathecal apertures in furrow between 4th and 5th segments, rarely between 3rd and 4th also. Setæ styliform or hooked, not forked, straight or slightly sigmoid, without distinct node; generally in fan-like bundles of 3 to 12, rarely fewer: arranged in 4 bundles per segment in all British genera. Head pore present. Male apertures 1 pair, in 12th; female 1 pair, in 13th. Testes in 11th, on septum between 10th and 11th segments; ovaries in 12th on that between 11th and 12th.

genus *Marionina* Mchlsn. Setæ sigmoid. Head pore small between prostomium and 1st segment: no dorsal pores. Dorsal blood vessel starts from mid-gut plexus behind clitellum; no hearts. Testes moderate, vasa deferentia long. Blood yellow-red. Œsophagus without pouches in 6th segment; not sharply marked off from mid-gut.

M. crassa (Clap.). Pale red. Setæ 3 or 4, rarely 2 or 5 per group. Lymph corpuscles of two forms. Penis small, sub-spherical. Total segments 40 to 48. Length about 15 mm. Seashore. Hebrides (Skye).

M. ebudensis (Clap.). Yellowish. Segments about 47. Length 12 mm. Seashore. Hebrides (Skye).

genus *Lumbricillus* Örst. Characters as in *Marionina*, except testes composed of numerous pyriform lobes.

L. verrucosus (Clap.). Reddish. Setæ 3 to 5 per group. Total segments 45 to 50. Length 10 to 12 mm.: thickness 0·5 mm. Seashore. Hebrides (Skye).

genus *Enchytræus* Henle, em. Mchlsn. Setæ straight, of equal length in same group. Head pore, dorsal pores, blood vessel, hearts, and œsophagus as in *Marionina*. Blood colourless. Spermathecæ communicate with gut.

E. argenteus Mchlsn. Silvery white. Lymph corpuscles by transmitted light nearly black. Total segments 23 to 30. Length

2·5 to 5 mm.: thickness 0·2 mm. Banks of streams, under stones, etc. (Kew?)

E. pellucidus Friend. White, transparent. Lymph corpuscles by transmitted light transparent. Total segments 60. Length 19 mm. In manure.

genus *Fridericia* Mchlsn. Dorsal pores present. Setæ straight. Ampullæ of spermathecæ with two diverticula.

F. magna Friend. Preclitellar setæ 3 or 4 per group, postclitellar 2. About 6 annuli of gland cells per segment. Clitellum ½ 11th and 12th segments. Blood red. Total segments about 90. Length 35 to 40 mm. In damp places.

F. agricola J. P. Moore. Anterior setæ 4 (rarely 5) per group, posterior 2 (rarely 3). Segments 65. Length 20 to 25 mm. In earth.

Family HAPLOTAXIDÆ. Male apertures 2 pairs, in 12th or 11th and 12th. Female apertures 1 or 2 pairs in furrow between 12th and 13th, or between 13th and 14th also. Spermathecal apertures 1 to 3 pairs, between 6th and 7th or 7th and 8th or 8th and 9th. Testes 2 pairs, in 10th and 11th: vesiculæ seminales present. Ovaries 1 or 2 pairs in 12th, or 12th and 13th. Spermathecæ without diverticula. Setæ sigmoid, not forked, hooked.

genus *Haplotaxis* Hoffmstr. Male apertures in 11th and 12th. Female between 12th and 13th, and 13th and 14th. Dorsal and ventral setæ not of equal length.

H. gordioides (G. L. Hartm.). Reddish. Setæ solitary, 2 or 4 per segment; dorsal setæ shorter than ventral, only present on variable number of anterior segments or absent. Prostomium sugar-loaf shaped, often separated by a more or less distinct circular furrow. First segment very short. Length up to 300 mm. Thickness 1·1 mm. In swamps, ditches, springs, etc.

Family GLOSSOSCOLECIDÆ. Setæ sigmoid, generally not forked, 8 per segment. No dorsal pores. Clitellum generally beginning behind 14th segment. Male apertures in or in front of clitellum, only exceptionally behind it.

Subfamily CRIODRILINÆ. Gizzard rudimentary.

genus *Sparganophilus* Benham. Prostomium not marked off by furrow from 1st segment. Male apertures not on raised lobes, between 18th and 19th or in anterior part of 19th. No œsophageal pouches or calcareous glands. Posterior hearts in 11th. 2 pairs of testes, 2 pairs of racemose vesiculæ seminales in 11th and 12th. Setæ closely paired.

S. tamesis Benham. Red, bluish iridescence. 3 pairs of spermathecal apertures, between 6th and 7th, 7th and 8th, 8th and 9th in line of dorsal pairs of setæ. Dorsal pairs of setæ not halfway up side of body. Nephridiopores in line of inner setæ of ventral pair. Clitellum from (part of) 15th to (part of) 25th segment. Tubercula pubertatis on 17th to 22nd. Male apertures between 18th and 19th. Length 76 to 102 mm. On roots of water plants. Thames.

Family LUMBRICIDÆ. Setæ sigmoid, not forked, 8 per segment. Dorsal pores present. Clitellum beginning at some distance from male aperture. Male apertures in 15th, female in 14th, usually. Genital and grooved setæ on papillæ developed on certain anterior segments. Œsophagus with calcareous glands. Gizzard well developed. Testes 2 pairs, in 10th and 11th. Ovaries in 13th. Spermathecæ simple, without diverticula.

genus *Eiseniella* Mchlsn. (*Allurus* Eisen). Prostomium extends about ⅓ across 1st segment. Gizzard confined to one segment. Spermathecal apertures dorsal to line of dorsal pairs of setæ.

E. tetraedra (Sav.), var. *tetragonura* (Friend). Sienna coloured; clitellum dull orange. Clitellum extends over 5 segments, from 18th to 22nd; tubercula pubertatis on 3 segments, from 19th to 21st. Male apertures on 13th segment. Total segments about 85. Length 30 mm. In gravel at bottom of streams, especially such as are occasionally dried up.

E. macrura (Friend). Greenish. Clitellum from 15th to 22nd. Tubercula pubertatis on 20th and 21st. Male apertures on 13th, ventral papillæ on 13th and 22nd. Total segments 160. Length 30 mm.

genus *Eisenia* Malm, em. Mchlsn. (part of *Allolobophora* Eisen). Prostomium dents or extends a little way into 1st segment. Male apertures in 15th. Spermathecal apertures 2 pairs, between 9th and 10th, 10th and 11th, in line of dorsal setæ, or dorsal to it. 3 or 4 pairs of vesiculæ seminales in 9th to 12th or absent from 10th. Gizzard extends through more than one segment.

E. fœtida (Sav.). The "brandling." Red with purple or brown rings. Prostomium extends into 1st segment. Setæ closely paired. Clitellum from 24th, 25th or 26th (usually) to 32nd. Tubercula pubertatis on 28th (or ½ 28th) to 30th or 31st. First dorsal pore between 4th and 5th. Total segments 80 to 110. Length 60 to 90 mm. Thickness 3 to 4 mm. In dung-heaps and rich soil.

E. veneta (Rosa.), var. *hibernica* (Friend). Clear rosy red. Prostomium extends some way into 1st segment. Setæ widely paired. Clitellum from 27th to 33rd broad and flat. Tubercula pubertatis on 30th and 31st. First dorsal pore between 5th and 6th. Total segments 100 to 115. Length 35 to 45 mm. Thickness 3 to 4 mm. In Ireland.

E. rosea (Sav.). Blood red. Prostomium extends some way into 1st segment. Setæ closely paired. Clitellum from 24th, 25th, or 26th to 31st, 32nd, or 33rd. Tubercula pubertatis on 29th, 30th and 31st. First dorsal pore between 4th and 5th. Genital setæ on 9th or 10th and (or) 12th or 13th, sometimes 24th. Total segments 120 to 150. Length 25 to 60 mm. Thickness 3 to 4 mm. In damp earth, and in mud at edges of fresh water.

genus *Helodrilus* Hoffmstr., em. Mchlsn. Prostomium generally extends into 1st segment, sometimes only dents it, sometimes extends right back to furrow between 1st and 2nd segments. Male apertures on 15th. Gizzard extends through more than 1 segment.

subgenus *Allolobophora* Eisen., em. Rosa. Prostomium does not reach furrow between 1st and 2nd. 4 pairs of vesiculæ seminales in 9th to 12th, those in 10th nearly as large as those in 9th. Setæ fairly closely paired.

H. (*A.*) *georgii* (Mchlsn.). Colourless. Prostomium extends about ⅓ across 1st segment. Clitellum from 28th or 29th to 35th. Tubercula pubertatis on 31st and 33rd. First dorsal pore between 4th and 5th. Total segments 105 to 110. Length 24 to 29 mm. Thickness 2·5 mm. Ireland.

H. (*A.*) *caliginosus* (Sav.). Colour very variable, but not purple. Prostomium as in *H. georgii*. Clitellum saddle-shaped, from 27th or 28th to 34th or 35th. Tubercula pubertatis on 31st and 33rd, or confluent from 31st to 33rd. Genital setæ on 9th, 10th and 11th. Vesiculæ seminales in 9th and 10th small. First dorsal pore between 9th and 10th, or rarely 8th and 9th. Total segments 104 to 248. Length 60 to 160 mm. Thickness 4 to 5 mm. In cultivated soil.

H. (*A.*) *longus* (Ude.), (*A. terrestris* Rosa). Smoky grey, iridescent. Prostomium as in preceding. Clitellum from 27th or 28th to 35th. Tubercula pubertatis from 32nd to 34th. Genital setæ on 9th, 10th, 11th, 31st, 33rd and 34th: those on last 3 named on distinct papillæ. Vesiculæ seminales of 9th and 10th small. First dorsal pore between 12th and 13th. Total segments 160 to 200. Length 120 to 160 mm. Thickness 6 to 8 mm. In cultivated soil.

H. (*A.*) *chloroticus* (Sav.). Colour variable, yellow, green, reddish. Prostomium extends ½ across first segment. Clitellum generally from 29th to 37th, rarely from 28th. Tubercula pubertatis small, on 31st, 33rd, 35th. Spermathecæ 3 pairs, between 8th and 9th, 9th and 10th, 10th and 11th. First dorsal pore between 4th and 5th. Total segments 80 to 125. Length 50 to 70 mm. Thickness 4 to 5 mm.

subgenus *Dendrobæna* Eisen, em. Rosa. Skin generally pigmented red. Prostomium generally extends into 1st segment, sometimes reaches across to touch furrow between 1st and 2nd. Setæ widely paired, or not paired. Spermathecæ 2 pairs, between 9th and 10th, and 10th and 11th: occasionally 1 or 2 more pairs on neighbouring segments. Vesiculæ seminales 3 pairs, in 9th, 11th and 12th; a 4th very small pair in 10th in species with widely paired setæ.

H. (*D.*) *rubidus*, var. *subrubicunda* (Eisen). Red dorsally. Prostomium extends $\frac{2}{3}$ into 1st segment. Body somewhat flattened. Setæ widely paired. Clitellum from 25th or 26th to 31st or 32nd. Tubercula pubertatis from 28th to 30th, confluent. Genital setæ on 16th. First dorsal pore between 5th and 6th. Total segments 60 to 110. Length 65 to 90 mm. Thickness about 4 mm.

H. (*D.*) *mammalis* (Sav.). Pale violet dorsally. Prostomium extends $\frac{1}{2}$ through 1st segment. Setæ separate, not paired. Clitellum from 31st to 36th segment. Tubercula pubertatis on 33rd and 34th, confluent. Total segments 98 to 100. Length 35 to 40 mm. Thickness 2 to 3 mm.

H. (*D.*) *octaedrus* (Sav.). Violet-brown, coppery. Prostomium extends $\frac{2}{3}$ into 1st segment. Setæ separate, not paired. Clitellum from 27th, 28th, or 29th to 33rd or 34th. Tubercula pubertatis from 31st to 33rd, confluent. Spermathecal apertures 3 pairs, between 9th and 10th, 10th and 11th, 11th and 12th. First dorsal pore between 4th and 5th. Total segments 80 to 95. Length 25 to 40 mm. Thickness 3 to 4 mm.

subgenus *Bimastus* H. F. Moore. Clitellum not reaching past 32nd segment. Tubercula pubertatis absent or inconspicuous. 2 pairs of vesiculæ seminales. No spermathecæ. Generally small and pigmented reddish.

H. (*B.*) *eiseni* (Levins). Pale violet dorsally. Prostomium reaches furrow between 1st and 2nd segment. Setæ closely paired. Clitellum from 24th or 25th to 32nd. No tubercula pubertatis. First dorsal pore between 5th and 6th. Total segments 75 to 110. Length 30 to 48 mm. Thickness 2 to 4 mm.

H. (*B.*) *constrictus* (Rosa). Red dorsally, especially anteriorly. Prostomium extends $\frac{2}{3}$ into 1st segment. Setæ widely paired. Clitellum from 26th to 31st. No tubercula pubertatis. Setæ on 16th on broad papillæ. First dorsal pore between 5th and 6th. Total segments 90 to 105. Length 20 to 30 mm. Thickness 3 mm.

genus *Octolasium* Örley, em. Rosa (part of *Allolobophora* Rosa). Prostomium extends into 1st segment, sometimes crosses it.

Testes and seminal funnels in paired vesicles, or in incompletely shut up portions of cœlom. More than 2 pairs of spermathecæ. 4 pairs of vesiculæ seminales in 9th to 12th. Tubercula pubertatis confluent. Gizzard extends through more than 1 segment.

O. lacteum (Örley), (*A. cyanea* and *A. rubida* Rosa). Bluish grey. Prostomium extends $\frac{1}{3}$ to $\frac{2}{3}$ into 1st segment. Setæ widely paired to separate. Clitellum from 30th to 35th. Tubercula from 31st to 34th, or with part of 30th and 35th. First dorsal pore between 8th and 9th, or 9th and 10th, or 10th and 11th. Total segments 100 to 165. Length 40 to 100 mm. Thickness 3 to 5 mm.

genus *Lumbricus* L., em. Eisen. Prostomium always extends across 1st segment and meets furrow between 1st and 2nd. Setæ closely paired. Clitellum saddle-shaped. Tubercula pubertatis confluent. Testes and seminal funnels in a single unpaired vesicle in 10th and 11th. 2 pairs of spermathecæ. 3 pairs of vesiculæ seminales, in 9th, 11th and 12th. Gizzard extending through more than 1 segment.

L. rubellus Hoffmstr. Red-brown to violet dorsally: feebly iridescent. Clitellum from 26th or 27th to 32nd. Tubercula pubertatis from 28th to 30th and sometimes 31st. First dorsal pore between 7th and 8th. Male apertures inconspicuous. Total segments 95 to 150. Length 70 to 150 mm. Thickness 4 to 6 mm.

L. castaneus (Sav.). Chestnut brown to brown-violet dorsally: iridescent. Clitellum from 28th to 33rd. Tubercula pubertatis from 29th to 32nd. First dorsal pore between 6th and 7th. Male aperture rather inconspicuous. Total segments about 90. Length 30 to 50 mm. Thickness 4 mm.

L. terrestris L. Müll. (*L. herculeus*, Rosa). Brown-violet dorsally anteriorly with darker mid-dorsal line posteriorly. Hinder end flattened. Setæ at extremities larger than in mid-region and not so closely paired. Clitellum from 31st or 32nd to 37th. Tubercula from 33rd (or $\frac{1}{2}$ 33rd rarely) to 36th (or $\frac{1}{2}$ 36 rarely). Male apertures conspicuous. Genital setæ generally on 26th, rarely on 25th also. First dorsal pore between 7th and 8th. Total segments 110 to 180. Length 90 to 300 mm. Thickness 6 to 9 mm.

L. papillosus Friend. Pale red-brown. Hinder end flattened. Setæ at extremities larger than in mid-region and not so closely paired. Clitellum from 33rd to 37th or ½ 38th. Tubercula pubertatis from 34th to 37th, papilliform on 34th and 36th. Male apertures conspicuous. Genital setæ on 29th to 32nd and 38th and 39th, or some of these. First dorsal pore between 9th and 10th, or rudimentary (?) up to between 6th and 7th. Total segments about 130. Length about 100 mm. Thickness (max.) 8 mm.

L. festivus (Sav.). Light red-brown. Clitellum from 34th to 39th. Tubercula pubertatis from 35th to 38th. Male apertures conspicuous. First dorsal pore between 5th and 6th. Total segments 100 to 120. Length 55 to 100 mm. Thickness 5 mm.

TABLE OF LUMBRICIDÆ (BRITISH).

	Clitellar segments	Tubercula pubertatis on segments	Male aperture in segment	First dorsal pore between segments	Setæ
Eiseniella tetraedra	18—22	19—21	13	4—5	Closely paired
,, macrura	15—22	20, 21	13	4—5	,, ,,
Eisenia fœtida	24, 25 or 26	28 or ½ 28—30 or 31	15	4—5	,, ,,
,, veneta	27—33	30, 31	15	5—6	Widely paired
,, rosea	24, 25 or 26—31, 32 or 33	29 or ½ 29, 30, 31	15	4—5	Closely paired
Helodrilus (Allolobophora) georgii	28 or 29—35	31, 33	15	4—5	,, ,,
,, ,, caliginosus	27 or 28—34	31, 33	15	9—10	,, ,,
,, ,, longus	27 or 28—35	32—34	15	12—13	,, ,,
,, ,, chloroticus	29—37	31, 33, 35	15	4—5	,, ,,
,, (Dendrobæna) rubidus	25, 26 or 27—31 or 32	28 or 29—30	15	5—6	Widely paired
,, ,, mammalis	31—36	33, 34	15	?	Not paired
,, ,, octaedrus	27, 28 or 29—33 or 34	31—33	15	4—5	,, ,,
,, (Bimastus) eiseni	24 or 25—32	none	15	5—6	Closely paired
,, ,, constrictus	26—31	28, 29	15	5—6	Widely paired
Octolasium lacteum	30—35	part 30 or 31—34 or part 35	15	8—9 or 9—10 or 10—11	Widely to not paired
Lumbricus rubellus	26 or 27—32	28—31	15	7—8	Closely paired, ventral less so
,, castaneus	28—33	29—32	15	6—7	Closely paired, ventral very slightly less so
,, terrestris	31 or 32—37	33 or ½ 33—36 or ½ 36	15	7—8	Closely in mid-region, widely at extremities
,, papillosus	33—37 or ½ 38	34—37	15	9—10 (rud. to 6—7)	,, ,, ,,
,, festivus	34—39	35—38	15	5—6	Closely paired

CHAPTER II.

THE CRAYFISH.

(*Astacus torrentium.*)

THE river Crayfish, *Astacus torrentium,* occurs in many English streams, especially those draining districts with calcareous soil. The species most usually obtained in laboratories is the continental *A. fluviatilis,* distinguished by the red tips to its claws. The former is abundant in many small streams on the borders of Surrey and Hampshire, and is still to be found in some small tributaries of the Thames, though its numbers have not yet recovered from the disease that in 1887 caused great mortality. Previous to this epidemic they were so abundant in the Thames and its tributaries that a regular industry of making crayfish-pots flourished in many places. The pots were baited with any carrion that came handy. A ready market is always obtainable, for when boiled in salt water they form brilliant garnishing to salads and are in flavour little if at all inferior to large prawns. The crayfish most commonly used for culinary purposes on account of its superior flavour and appearance is *A. fluviatilis.* The supply is obtained from fresh-water lakes in central Europe and Russia. The species has become so scarce in Western Europe as to no longer furnish an industry, and is, apparently, being steadily pushed (or exterminated) eastwards. The contents of the lakes are put up to auction under the Russian Government. The lessee keeps a fleet of fishers, and further pays the men so much per 60 crayfish. The marketable specimens are brought westwards by easy stages, and at each halting place are put in streams for a week to recover from the effects of their journey. Between London and Russia there are three of these stations of refreshment. The animals fetch from twelve to as much as sixty shillings per hundred (wholesale prices) according to the time of year and size and condition. *A. torrentium* is also found in some lakes and pools and is stated to have been taken some distance out at sea. Its length is about 3—3½ inches, and in colour it is a dull greenish-grey, being thus admirably concealed among its natural surroundings. During the daytime it usually remains in holes which it digs in the bank of the stream, generally from 3 to 15 inches below the water level, and lies just at the mouth of its burrow with the nippers and long antennæ projecting out as sentinels into the open water. It is very averse to strong light, and in captivity always selects a well-shaded nook of the aquarium. The food is very varied, living or dead, animal or plant being alike acceptable: worms are a convenient food for specimens in captivity, but they are also very partial to certain Algæ (*Chara*) and thickened succulent roots such as those of the carrot. The human finger, gloved or bare, is readily

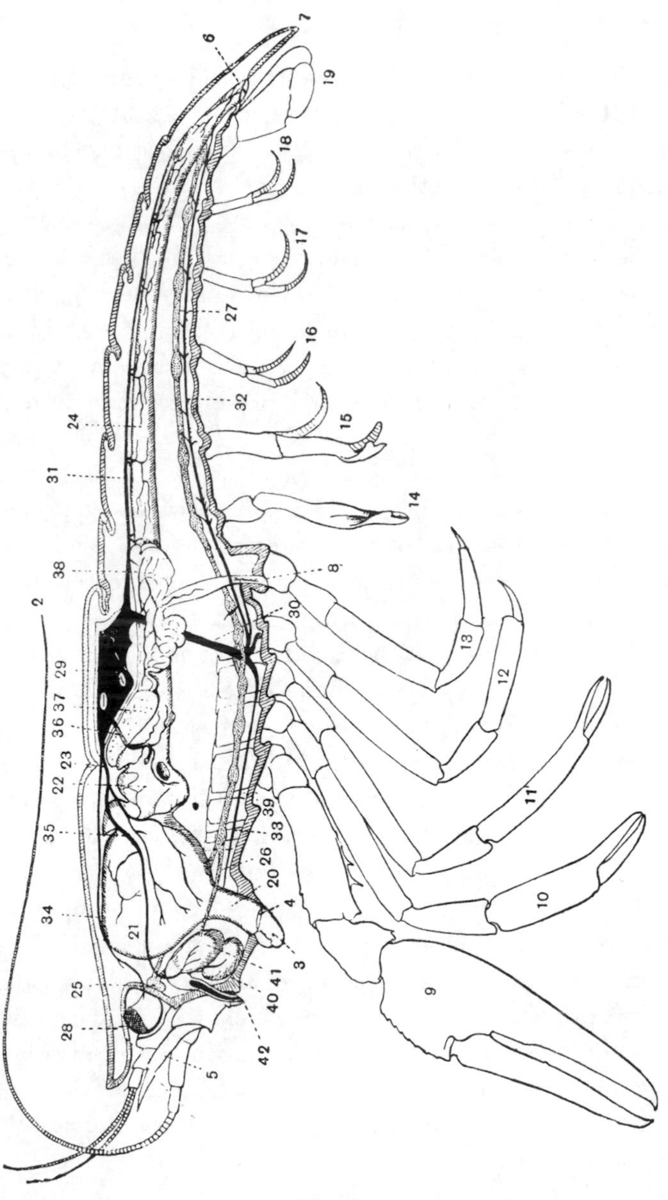

2
3
4
5
6
7
8
9
10
11
12
13
14
15
16
17
18
19
20
21
22
23
24
25
26
27
28
29
30
31
32
33
34
35
36
37
38
39
40
41
42

Fig. 7.

Fig. 7. Semi-diagrammatic view of internal organs, and some limbs of right side of a male crayfish, *Astacus fluviatilis* × 1. Partly from Howes.

1. Antennule. 2. Antenna. 3. Mandible. 4. Mouth. 5. Scale or squama of antenna, exopodite. 6. Anus. 7. Telson. 8. Opening of vas deferens. 9. Chela. 10. 1st walking leg. 11. 2nd walking leg. 12. 3rd walking leg. 13. 4th walking leg. 14. 1st abdominal leg, modified. 15. 2nd abdominal leg, slightly modified. 16. 3rd abdominal leg. 17. 4th abdominal leg. 18. 5th abdominal leg. 19. 6th abdominal leg, forming with telson the swimming paddle. 20. Œsophagus. 21. Stomach. 22. Mesenteron, mid-gut. 23. Cervical groove. 24. Intestine. 25. Cerebral ganglion. 26. Para-œsophageal cords. 27. Ventral nerve-cord. 28. Eye. 29. Heart. 30. Sternal artery. 31. Supra-intestinal artery. 32. Sub-intestinal artery in abdomen. 33. Sub-intestinal artery in thorax. 34. Ophthalmic artery. 35. Antennary artery. 36. Hepatic artery. 37. Testis. 38. Vas deferens. 39. Internal skeleton. 40. Green gland. 41. Bladder. 42. External opening of green gland.

seized if carefully insinuated into the mouth of the burrow, and may be conveniently used as a bait by the amateur crayfish hunter.

When advancing in search of food the animal moves stealthily forward, walking slowly by means of its thoracic legs assisted by the paddling action of the abdominal swimmerets. In walking, the first three pairs of legs pull and the fourth pair pushes. Their order of movement is as follows: the first on the right and the third on the left side move together, next the third right and first left, then the second right and fourth left, and lastly the fourth right and second left. The body is thus always supported by six feet while the other two are advancing[1]. The nippers (*Chelæ*) are usually carried clear of the bottom, and partly flexed so as to be capable of quick movement forward. The third maxillipedes search the bottom in a tentative manner and appear to possess a very delicate

[1] List, *Morph. Jahrb.* XXII.

sense of touch (taste?). It is probable that the fine bristles with which they are fringed are sensory in function. If inanimate food is found by these appendages they at once rake it back to the mandibles. Should the food be some living animal it is seized and held firmly by one or both nippers, and by them and the small nippers torn up and ultimately conveyed to the mouth, where the powerful mandibles cut off pieces which are swallowed and pass up into the 'stomach' for further mastication. When alarmed the Crayfish moves with alacrity: bringing into play its broad tail-fin and telson; bending the abdomen repeatedly downwards and forwards by a few vigorous contractions of the powerful flexor muscles which occupy the greater part of the abdominal cavity, it darts swiftly backward into some sheltering recess. It is however unable to travel more than a few yards at a time in this way, the muscles soon becoming exhausted. Experiments[1] have shown that under electrical stimulus the contraction of the abdominal flexor muscles only lasts one-tenth of the time occupied by that of the muscles of the great chelæ; from 80 to 100 stimulations per second are required to produce tetanus of the former, whereas from two to four suffice to bring about this condition in the latter. The tetanus of the flexors only lasts about half-a-minute, but that of the chelæ muscles nearly half-an-hour, and during the first five minutes of the period steadily increases in strength. The physiological properties of these muscles are thus exquisitely suited to the nature of the work performed by them. A crayfish will yield its life rather

[1] Richet, *vide Nature*, xx. p. 106, 1879.

than let go of its prey when once seized by the chelæ. This tenacity is of great value to it in wearing down the strength of a struggling opponent: on the other hand, a few spasmodic jerks of its abdomen suffice, as a rule, to take the animal to a place of safety.

Food and Digestion. The nature of the food has already been mentioned. On reaching the 'stomach' it undergoes preparation for the action of the digestive fluids to which it is subsequently exposed. The 'stomach' contains a number of ossicles, three of which carry teeth projecting into the cavity. These teeth are caused to clash together and thus triturate the food by the action of muscles, some of which are extrinsic and attached to the outer covering of the body, while others lie in the wall of the 'stomach' itself. At the hinder end of the organ is a sieve of crossed bristles which allows only finely divided matter to pass through. The action of this 'gastric mill' may be most easily understood by constructing a cardboard model as follows:

Cut out a piece of card shaped as in Fig. A. Along *ab, cd, ef, hi,* and *mn* cut just the surface of the card with a penknife; do the same, but on the opposite face of the card, along *gk* and *lo*. Then bend slightly downwards the triangular pieces 2, 2; turn 9, 9 under the piece 6, 5, 6 until the lower surfaces of 9, 9 are flat against that of 6, 5, 6: stitch the shaded part of 9, 9 firmly by thread or fine wire to 6, 5, 6; then bend the unshaded part of 9, 9 till at right angles to the shaded part, using *lo* as hinge-line. These projecting pieces of 9, 9 then represent the lateral teeth.

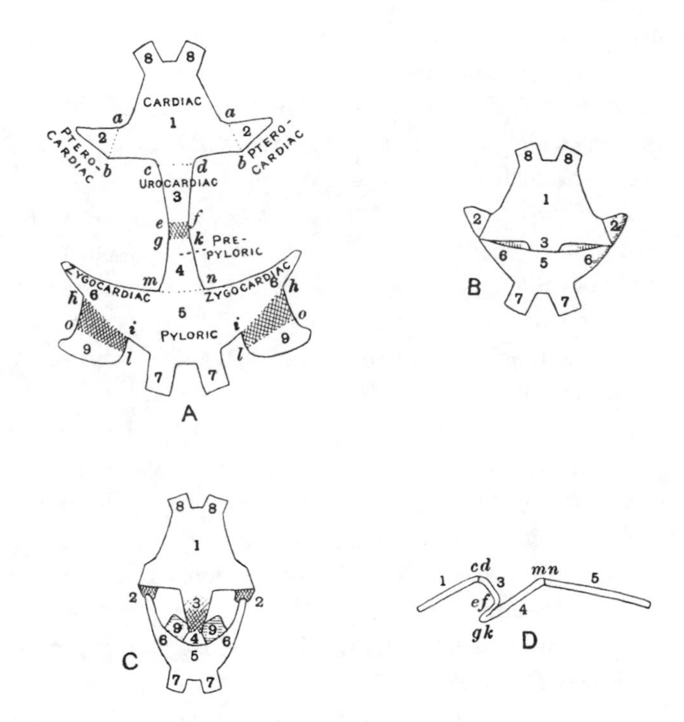

A. Cardboard as first cut out. B. Model complete : at rest. C. Model complete : muscles contracted. D. Median vertical section of model to show folds. After W. E. Roth, *Nature*, vol. XXI. Feb. 26th, 1880.

For explanation of figures and letters see the text.

Next bend the piece 1, 3, 4 upon hinge-line *gk*, until the shaded portion is flat upon the surface of 4, where it must be securely stitched; this done bend back 1, 3 on hinge-line *ef* until 3 is at right angles to 4. The projecting end of 4 made prominent by these folds represents the central tooth. The piece 1 must now be bent gently downwards upon 3, using *cd* as hinge-line, and 4 must be bent sharply on 5, using *mn* as hinge-line. Lastly, perforate the corner of 6, 6 and of 2, 2, and by a single wire (to allow a certain amount of rotation) unite right hand 2 to right hand 6, and left hand 2 to left hand 6, in each case 2 being outside 6. To do this 6, 5, 6 must be bent like a bow, its right and left arms being thrust downwards and inwards. The model will then be as in Fig. B.

If now the pieces 8, 8 and 7, 7, which represent the anterior and posterior gastric muscles, are pulled so as to represent the effect of a muscular contraction the three teeth come sharply together, but are separated again and the whole model brought back to its original condition of the elasticity of the cardboard. Of course in the actual stomach of the crayfish the gaps between the ossicles (*vide* Fig. C) are filled in with thin, flexible chitin. By carefully adjusting the size and direction of the 3 teeth in the model and further by hardening them with sealing-wax or similar material, they may be made to grind bread, etc. into small fragments. A sectional view is shown in Fig. D.

The food in a state of fine division is exposed in the intestine to the action of the digestive fluid secreted by the so-called 'liver,' whose ducts open into the extremely short mesenteron. The secretion has an acid reaction and, in addition to ferments—analogous in their action to those produced by human salivary, gastric, and pancreatic glands—contains also a coloured substance called hæmatin, which is an organic compound of iron. It appears that the 'liver' is not only a digestive gland but is also concerned in the preparation and storage of pigment which is perhaps utilised in giving colour to the outer shell. Cuénot[1] has shown, by employing meat stained with various dyes, that the 'liver' tubules are concerned in the absorption of food, much of the coloured meat passing into them. This observation is confirmed by McMurrich's[2] work on terrestrial Isopod Crustaceans. Saint Hilaire[3] has proved that peptone is not absorbed by the intestine, and that if injected in small quantity into the body spaces it is taken up by the 'liver': in large quantity it causes death. Fats on the other hand are absorbed by the epithelium of the mid-gut and its dorsal diverticulum. The 'liver' is further of service in regulating the percentage of water in the blood, and in addition lays up stores of fat and glycogen. A structure known as the dorsal pyloric valve carries undigested and indigestible solids across the cavity of the mid-gut, whose epithelium is very delicate, direct into the intestine.

Respiration is carried on by means of the branchial plumes which project outward from the legs or sides of the thorax into the water, and are covered by a protecting fold of the shell (branchiostegite) so as to be practically enclosed in a chamber through which the water is wafted from behind forward in a steady flow by the sculling movements of the 'scaphognathite' (the united exopodite and epipodite of the second maxilla), at the anterior end of the branchial chamber. The apex of the endopodite of

[1] *C. R. Acad. Sci.* Paris, tom. 116.

[2] *Journ. Morph.* XIV. p. 102.

[3] *Bull. Acad. R. Belg.* (3) XXIV. p. 506; cf. also Stamati, *Bull. Soc. Zool. Fr.* XIII. p. 146.

this appendage hooks into a small recess upon the mandible near the base of the palp. When the scaphognathite is at work it pulls upon this hook at each stroke. The number of strokes per minute is about sixty. The first and second maxillipeds may be seen to be occasionally thrown into very rapid flickering vibrations. The purpose of these movements is not at all clear. When the animal is at rest the third maxillipeds, chelæ, and some of the ambulatory appendages are often kept in gentle movement, swaying to and fro from right to left. These actions doubtless assist in changing the water in the vicinity of the respiratory organs.

The branchial plumes contain blood which flows up the outer and down the inner side of the stem of each plume. The inner wall of the branchiostegite itself has been demonstrated by Claus[1] to be highly vascular. No doubt it plays an important part in respiration. The blood is almost colourless[2], and contains, dissolved in the liquid and not contained within any corpuscles, an organic compound of copper (hæmocyanin), which forms an unstable blue compound with oxygen as the blood courses through the branchiæ and readily parts with the loosely associated gas to the other tissues of the body when it reaches them. When shed the blood soon turns indigo blue in consequence of the oxidation of the hæmocyanin. It clots quickly in virtue of the action of a ferment, yielded by its amœboid corpuscles, upon a globulin proteid present in the plasma. A reddish tinge that may be observed in the blood and

[1] Bouvier, *C. R. Acad. Sci.* Paris, tom. 110.

[2] Halliburton, *Journ. of Physiol.* VI. No. 6, 1885.

also in the skeleton and hypodermis is due to 'tetronerythrin,' a pigment which also has a strong affinity for oxygen. The amœboid corpuscles are active in devouring bacteria and other injurious organisms that have obtained entry into the system.

The heart, which receives the blood through its valved ostia from the pericardial sinus, is stimulated to contract by the entrance of the oxygenated blood[1]. The hinder portion contracts first and then the anterior, by which time the hind section has begun to dilate. Rise of temperature causes the rate of beating to increase. The average rate is from sixty-five to eighty-five beats per minute. The heart is under the control of the nervous system, and has a double nerve-supply, (1) a median unpaired nerve from the stomatogastric ganglion, (2) nerves from the thoracic ganglia of the ventral chain: stimulation of the former quickens, but of the latter retards the heart-beat. The heart may be revived to activity after the beats have ceased, and may be maintained beating for some time by the application of weak acetic acid.

The excretion of waste nitrogenous matter is effected by the pair of green glands, whose discharging apertures are in the bases of the large antennæ. Uric acid and other nitrogenous compounds akin to urea have been detected in these glands (Griffiths).

Nervous System and Special Senses. The physiology of the nervous systems of Invertebrates has been investigated but little. It is however known that in the cray-

[1] Plateau, *vide Nature*, XIX. p. 470, 1879.

fish[1] there is no crossing of nerve fibres from right to left sides of the body and *vice versa*, and thus that the right portion of the central nervous system controls the right half of the body and similarly the left, *mutatis mutandis*. The brain (supra-œsophageal ganglia) is the seat of all voluntary as distinguished from reflex actions; it also is responsible for the maintenance of equilibrium and for the use of the abdomen in swimming, and moreover controls by inhibition the purposeless activity of the lower nerve centres. The subœsophageal ganglion mass coordinates the feeding and locomotor movements and sets up peculiar rhythmic swinging of the limbs if the brain is removed. In the absence of these first two ganglion masses all co-ordination is lost and the limbs move in an aimless way, even to the extent of interfering with each other as if not members of the same individual; for example, the chelæ will rob each other of food and generally play at cross purposes.

The vision of Crustaceans and of all Arthropods has given rise to much discussion and it is not our intention to go into this vexed question. But it may be pointed out that even though the compound eye may produce on the retinal cells a mosaic of numerous small pictures fitted one to another, it by no means follows that the impression received by the animal in any way corresponds with this mosaic. We ourselves, in common with all Vertebrates, receive through the optical media of our eye an inverted picture upon the retina, but the sensation produced is that of an erect and not inverted object.

[1] Ward, *Proc. Roy. Soc.* XXVIII. p. 379, and *Journ. Physiol.* II. p. 214.

In the basal joint of the first antenna is situated the 'auditory' sac (otocyst), whose aperture is protected with branching 'guard' setæ and whose curved floor bears two rows of hollow, almost simple, sensory setæ which are moveably articulated to the wall of the sac. Within the sac are numerous otoliths, which are chiefly, if not entirely, foreign bodies, such as grains of sand. These are scattered by the chelæ over the aperture of the sac and thus fall into it. It is highly probable that this organ is of more service as an instrument for informing the crayfish of its attitude and for the maintenance of the equilibrium of the body than as an organ of hearing. Martha Bunting[1] has proved experimentally that extirpation of both otocysts brings about a complete upset of the animal's sense of equilibrium. A crayfish thus mutilated will swim about upon its back for a long while apparently unconscious of the inverted posture.

Kreidl[2] by a beautiful experiment, which in this country has not received the attention that it deserves, has indeed placed the matter beyond the region of dispute. Remembering that at each moult the otoliths (statoliths) are removed with the cast-off cuticle, and that fresh foreign particles are then put in by the animal, Kreidl conceived the idea of causing some force other than that of gravity to act upon them. Finely powdered particles of iron were accordingly supplied to specimens of *Palæmon xiphios* and *P. squilla* (marine Decapods allied to the crayfish)

[1] *Pflüger's Archiv f. d. ges. Physiol.* LIV. p. 531.

[2] *Sitz. Wiener Ak. d. Wiss.* CII. 1893, Abth. 3; and cf. Hensen, *Zeit. wiss. Zool.* XIII.

that had just moulted. The animals were kept in water carefully filtered so as to exclude all other solid particles. When they had been observed to place iron particles in their otocysts (statocysts) an electro-magnet was brought to bear upon them. (This instrument is preferable to a permanent magnet, for it can be placed in the desired position near the tank while the electric circuit is open. On closing the circuit it at once becomes a magnet and the behaviour of the animals can be noticed without the possibility of disturbances and excitement that might arise from the movements necessary in bringing a permanent magnet within range.) Under these conditions the animals adjusted the positions of their bodies in obedience to the influence of the magnet upon the particles of iron. In the normal positions the plane (*BA*, *BC*, Fig. 8) in which

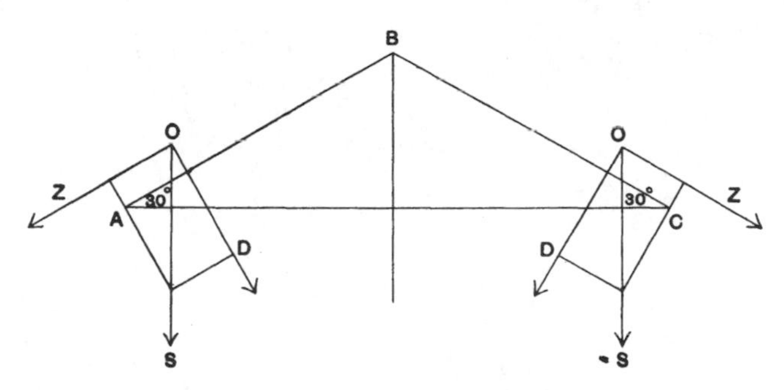

Fig. 8.

the statoliths lie upon the sensory setæ makes an angle of about 30° with the horizontal. Hence the force of gravity acting upon the particles can be resolved into two components; one which presses at right angles to the supports of the statoliths (viz. the setæ) which may be termed the 'pressure-component' (*OD*, Fig. 8), the other which exercises an outward pull upon the statoliths parallel to the planes *BA*, *BC*: this may be termed the 'pull-component' (*OZ*). The animal regulates its position in accordance with the pressure-component.

When the magnetic force is brought into action it is, of course, in addition to the existing force of gravity and the total effect is, so to speak, a compromise (or alliance) between the two forces, or rather between their two pressure-components. Dealing now with the right side of the body only, in Fig. 9, where the magnet *M* is placed above and to the right of the animal, *OS*, the direction of the force of gravity upon *O*, the centre of gravity of the

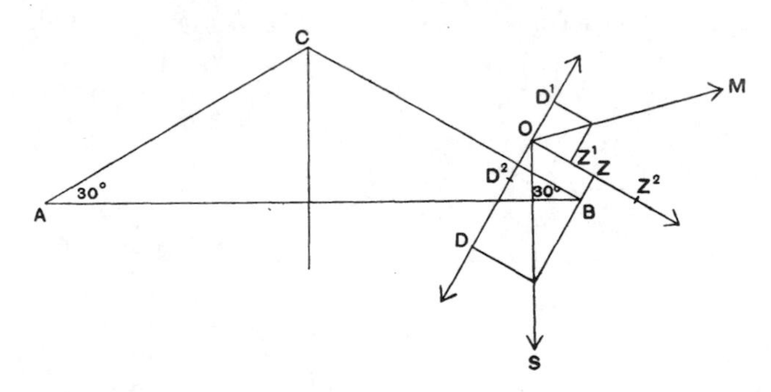

Fig. 9.

otolith (statolith) may be resolved into *OD* (pressure-component) and *OZ* (pull-component), while *OM*, the direction of magnetic attraction, may be resolved into OD^1 and OZ^1. It will be seen that OD^1 is in the contrary direction to *OD* and therefore diminishes the effect of the latter upon the sensory setæ. A similar diminution would be, and on all previous occasions within the experience of the animal has been, effected by the downward inclination of the right side of the animal's body. Hence the sensation

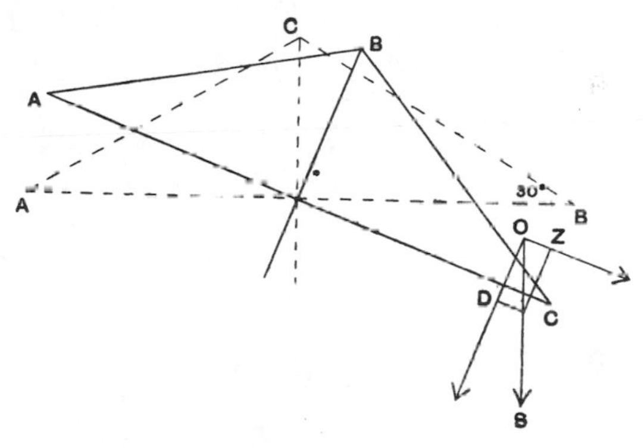

Fig. 10.

produced is that of slipping downwards to the right. Accordingly the creature inclines itself towards the right as though walking on a surface sloping downwards in that direction (Fig. 10). Again, if, as in Fig. 11, the magnet

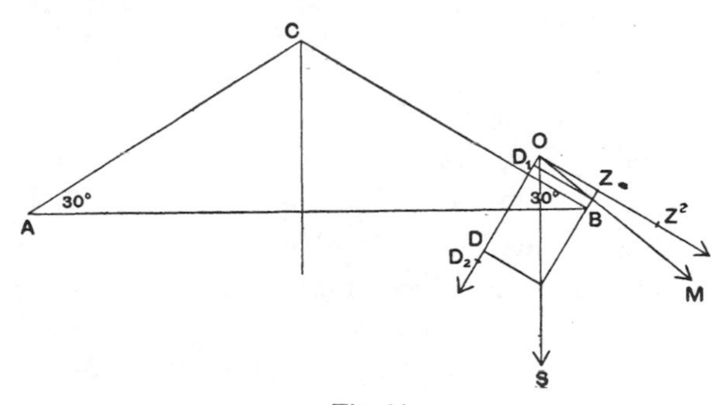

Fig. 11.

be placed to the right but below the animal, then OD^1 and OD are in the same direction and reinforce each other, producing an increased pressure OD^2 upon the setæ. The sensation now caused is that of slipping down to the left and the animal adjusts its body accordingly (Fig. 12). Kreidl offers no explanation of the effect

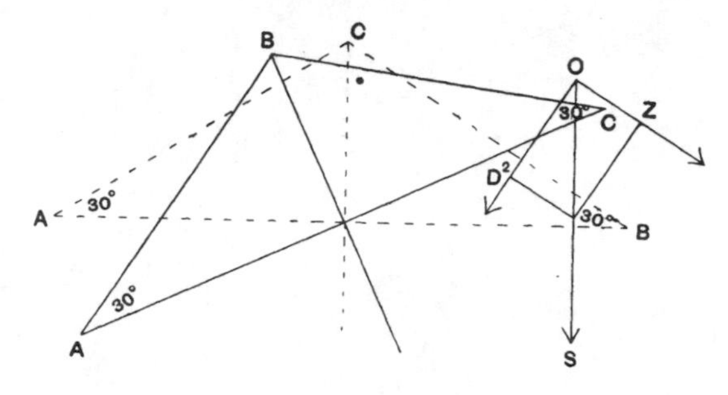

Fig. 12.

of the pull-components OZ, OZ^1. I have somewhat modified the mechanical presentation of the argument as expressed by the deviser of the experiments, but believe that the main substance of the reasoning is identical.

It would be interesting to try the effect of placing the magnet vertically above the animal, a position apparently not selected by Kreidl. If the magnet were of sufficient strength to entirely neutralise the force of gravity and draw the iron particles upward, one would expect the animal to 'turn turtle' and remain inverted so long as the magnetic force persisted. The terms 'statocyst' and 'statolith,' suggested by Kreidl in place of 'otocyst' and 'otolith' respectively, are far preferable, and may with advantage be extended to similar structures in other invertebrate and vertebrate animals.

True, it has been shown that the "auditory" setæ of some Crustaceans are thrown into vibration by certain noises and musical notes, but this does not prove that they give rise to the sensation of hearing. A loud noise will throw many of the wires of a piano into vibrations and cause them to give out their respective notes. Similarly these "auditory" setæ will undoubtedly respond to their appropriate notes, but it seems unreasonable to infer on that account that they are sense organs for the perception of waves of sound.

Crayfish undoubtedly possess a sense of smell: this sense has been attributed to the flattened spoon-shaped setæ found on the ventral surfaces of the more distal joints of the outer branch of the first antenna. It is interesting in this connexion to note that when the animal is at rest or slowly advancing the outer branch is always held up, pointing obliquely forward, and from time to time is jerked downwards and up again. The inner branch is held horizontally[1].

It is probable that the setæ which occur scattered over the general surface of the body and especially upon the antennæ are sensory (tactile ?) in function.

Reproduction, Development, and Regeneration. In September or October the ventral surface of the abdomen of the female crayfish assumes a white appearance due to the activity of certain cement glands, situated in the

[1] Cf. Boas, *Morph. Jahrb.* VIII. p. 490; Jourdain, *C. R. Acad. Sci.* Paris, XCI. p. 1091; and *Jour. de l'Anat.* XVII.; Leydig, *Müller's Arch.* 1860.

sternal portions of the anterior segments. These produce a secretion whereby the eggs are glued to the parent's body. The eggs are laid in November or December, and at this period the female bends her abdomen forward, somewhat into the position normal to that region in crabs, so as to enclose a chamber whose dorsal and ventral margins become fastened together by the above-mentioned cement. The eggs are received from the oviduct into this chamber and become attached by the cement to the fringes of the swimmerets and other parts of the ventral surfaces of the abdominal segments. Each egg has a short stalk of attachment. Impregnation is effected by the male throwing the female on her back and then pouring the spermatozoa through the tubular appendages of his first abdominal segment on to the margins of the oviducal apertures and on surrounding regions. The second pair of abdominal appendages of the male are during this process worked to and fro in the cavities of the first, as though thrusting the seminal fluid through them, or endeavouring to keep them clear. The female with eggs attached is popularly said to be "in berry." The water around the developing young is renewed, and aeration maintained by gentle swaying movements of the swimmerets.

The young crayfishes are hatched during the first half of the ensuing summer, and, as so frequently in fresh-water animals, are already far advanced in development, having undergone at least one moult of the skin while within the eggshell. In general form the young is not strikingly different from the adult. The cephalothorax is larger in proportion to the abdomen; the chelæ are more slender, their tips are more sharply curved inwards, and there are hooks on the ends of the last two walking legs. These last modifications are of value in enabling the young animal to cling to the empty eggshell or to the swimmerets of its parent for a time. In about 10 days it moults, and again four times at intervals of about 3 weeks until September; the skin then obtained lasts through the winter, but in the following spring and early summer three more moults occur, making 8 moults during the first year of life. During the 2nd year there are 5 moults, 2 in the 3rd. The female is mature in her fourth year, and thenceforth moults only once yearly. The male is mature in his 3rd year, and is said to continue moulting twice every year. It is not known to what age crayfishes attain, nor at what age growth ceases. It must be borne in mind that moulting may well continue after growth has ceased; the casting-off of the skin is an act of excretion, and probably accounts for the singularly feeble development of nitrogenous excretory organs in the crayfish and other Arthropods. The need for nitrogenous excretion does not cease when growth ceases, though probably the moults become less frequent. We need exact observation on this point.

The phenomena preceding a moult are complex, and affect organs other than the mere cuticle that is to be cast off. About 5 weeks before a moult, in an adult, or for a shorter time in younger individuals, the wall of the stomach on each side of the entrance of the œsophagus begins to lay down sheets of calcified chitin which eventually give rise to smooth, flat, and slightly concave stones, known as *gastroliths*[1], or popularly as "crabs' eyes" in crab and lobster. Before the moult actually occurs the stones are cast off into the cavity of the stomach, and are either ground down in the gastric mill or removed with the lining of the stomach when the moult takes place. These stones are chiefly composed of calcium carbonate, and it seems probable that their substance is derived from the old shell which is about to be cast off, and indeed that the lines along which the shell splits open are thus prepared and rendered weaker and less resistant by the removal of their mineral component which is then excreted by the wall of the stomach. When the time for the moult arrives the shell splits across the back at the junction of the thorax with the abdomen; other splits of varying magnitude and extent occur along the limbs. The crayfish, lying on its side, withdraws the cephalothorax and its limbs first, and then the abdomen, and remains in hiding and helpless for a few days until the new shell has hardened. It is interesting to note that stores of glycogen (a food substance allied to starch and sugar) are laid up in the connective tissues prior to the moult. Provision is thus made for the period of helplessness when the animal is unable to procure fresh food of any kind.

If frightened or injured a crayfish will often throw off one or more of its limbs, especially if the alarm is caused by one of these being seized: the separation takes

[1] *Vide* Irvine and Woodhead, *Proc. R. Soc. Edinb.* XVI. p. 330; Herrich, *Bull. U. S. Fish. Commis.* p. 93, 1895.

place at the junction of the first and second joints. At this spot a double diaphragm exists with a single central hole through which pass nerves and blood vessels. When the limb is cast off the hole, being very small, is soon stopped up by coagulated blood. From the stump a small new limb is developed, which, after a few moults, gains the normal size. The device is clearly for the protection of the crayfish: an enemy may seize a limb and keep it, but the owner escapes and is very little the worse.

Foes and Parasites. Crayfish fall victims to many birds of the Duck tribe, and to herons, who split them open by hard blows on the back. They are also caught and eaten by water-voles, being then dragged ashore and consumed on the bank.

Their parasites are not very noticeable. A very large Gregarine Protozoon, *Porospora gigantea*, is found in the alimentary canal. On continental crayfish, which are more often procurable from dealers than our English species, there occurs among the eggs and on the swimmerets of the female the remarkable leech (or Oligochæte worm?) *Astacobdella* or *Branchiobdella.* This leech is about ⅓ inch long, and has a well-developed body cavity divided by transverse partitions. Its eggs, which are more often to be seen than the adult creature, are laid singly and attached by a slender stalk to the swimmerets of *Astacus.* The parasite is charitably credited with devouring only the addled eggs of the crayfish, and thus acting the part of a useful scavenger.

Encysted upon the surface of the intestine, and elsewhere within the body, there are frequently to be found the encysted larvæ of an Acanthocephalan, *Echinorhynchus polymorphus*. This parasite becomes sexually mature in the intestines of ducks, swans, geese and other birds of similar habits. The adult male is about 5 mm. long, the female about 25 mm. Two species of Trematodes, *Distomum cirrigerum* and *D. isostomum*, are frequently found among the viscera. The former of these is usually in an encysted condition, the cysts varying from 0·2 mm. to 1·75 mm. in diameter. Zaddach[1] has observed self-impregnation occur in this species: the eggs are laid in the cyst; the parent dies and the eggs are scattered within the body of *Astacus* by the decay of the cyst. It is not known how the crayfish first becomes infected with the parasite.

Geographical Distribution. The genus *Astacus* is found in Europe and Western Asia as far south as the Aral and Caspian Seas, and also in America in the streams on the west of the Rocky Mountains.

[1] *Zool. Anz.* IV. 1881, pp. 398, 426.

CHAPTER III.

THE COCKROACH.

Periplaneta (Blatta) orientalis.

THE Cockroach or "Black-beetle," though now thoroughly established in England and known only too well as a household pest, is not a native of this country but of tropical Asia. It was introduced here, probably in the 16th century, and following the lines of commerce has now spread over the world. It is one of the most constant pests on board ship where it is said to gnaw the skin and toe-nails of the sailors. In some remote villages it is still unknown, and even to the naturalist Gilbert White it was in 1790 an "unusual insect." The nocturnal habits and retiring disposition of the animal are familiar. Its flattened and loosely jointed body is well suited for squeezing into narrow chinks and crevices, and the sombre colouring renders it the less conspicuous in its dark retreats. It is now so completely associated with human dwellings or places of industry, such as warehouses, dockyards, etc., and so seldom found out of doors that it is difficult to imagine what its habits may have been in pre-human periods.

The indigenous cockroaches, which are referred to below, live in shrubs, under moss, dried leaves, refuse heaps and so forth, and thus indicate the steps by which, in all likelihood, *P. orientalis* quitting a purely vegetarian diet came to gather up the unconsidered trifles cast out around primitive settlements, and so became intimately associated with man and man's doings. The factors which have contributed to the success of this animal are not far to seek; in addition to its form and colour already alluded to, it has a fair turn of speed and is difficult to catch, and when caught to hold, for its smooth surface readily slips from the grasp; and further it possesses a most disagreeable smell (and taste ?) which taints everything that it touches, and pervades its accustomed haunts. This evil odour is due to the secretion of a pair of glands situated on the back. They are sunk in the thin membrane which connects the 5th and 6th segments of the abdomen[1]. If a cockroach is caught by quickly putting the hand upon it as it runs over the floor, a sticky glue-like fluid is at once emitted by these glands and gives forth the characteristic odour with great intensity, nor is it at all easy to rid the hand of the taint. There can be little doubt that this quality is a sufficient deterrent to many insectivorous animals, though some few have overcome the dislike.

The male cockroach when adult possesses a pair of wings beneath the wing-covers by means of which flight

[1] Minchin, *Q.J.M.S*, xxix. and *Zool. Anz.* 326, 1890.

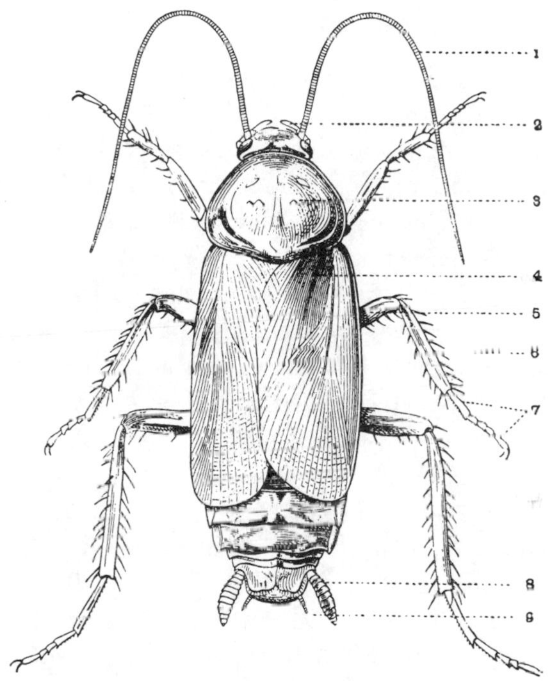

Fig. 13. *Periplaneta orientalis*, male × 2. Dorsal view. From Kükenthal.

1. Antenna. 2. Palp of first maxilla. 3. Prothorax. 4. Anterior wings. 5. Femur of second leg. 6. Tibia. 7. Tarsus. 8. Cerci anales. 9. Styles.

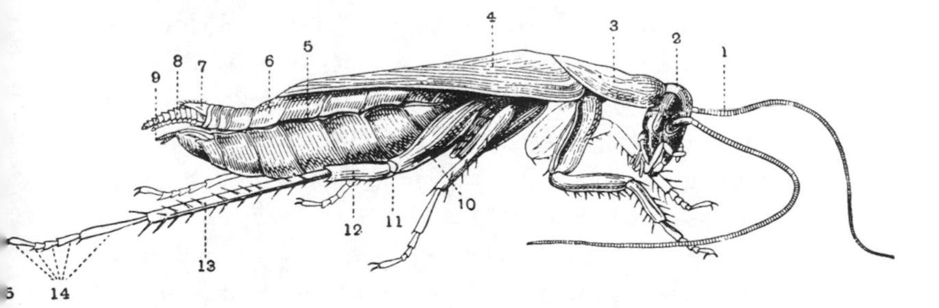

Fig. 14. *Periplaneta orientalis*, male × 2. Side view. From Kükenthal.

1. Antenna. 2. Head. 3. Prothorax. 4. Anterior wing. 5. Soft skin between terga and sterna. 6. Sixth abdominal tergum. 7. Split portion of tenth abdominal tergum. 8. Cerci anales. 9. Styles. 10. Coxa of third leg. 11. Trochanter. 12. Femur. 13. Tibia. 14. Tarsus. 15. Claws.

is possible, but it is a very uncommon occurrence to meet a specimen on the wing. In flight the wing-covers (anterior wings) are held almost at right angles to the body on the right and left sides and apparently make but slight active vibrations, acting chiefly as aero-planes. The posterior (true) wings are the organs of propulsion; their front edge is fairly strong and rigid and offers a cutting edge to the air and is not easily deflected, on the other hand the bellying hinder margin is far more pliant, and at each stroke of the wing is thrust upward by the pressure of the air so that its under surface slants obliquely and thus gives the insect a forward push. This action can be imitated by folding one edge of a half-sheet of ordinary note-paper so as to stiffen it for about half an inch all along and then, holding the sheet horizontally a few feet above the ground, release it and allow it to fall; the unfolded, more pliant portion of the paper will bend upwards a little and the sheet will not fall straight to the ground but will glide obliquely downwards, travelling steadily in the direction of the stiffened edge. In an experiment conducted with such a simple model a distance of about 9 feet was traversed horizontally by the sheet of paper in falling from a height of 7 feet. The propelling power of all wings whether of birds or of bats or of insects depends more or less upon a similar pliant hinder portion. The precise effect upon the flight of the varying arrangement of "veins" in insect wings is entirely unknown and presents a series of problems in ærostatics that would tax the highest mathematical ability.

When at rest the flight-wings of the cockroach are

folded. In arriving at the resting position two movements take place. First the posterior portion of the extended wing shuts up like a fan; then the anterior part, which remains flat and devoid of fan-like creases, moves back so as to cover the posterior section. Both the anterior and posterior free edges of the wing are now directed outwards laterally, while the creased edge formed by the main folding is turned towards the mid dorsal-line. The anterior firmer region of the wing lies uppermost so as to constitute a protection to the more delicate hinder, and now underlying portion. All the folds run longitudinally.

The legs are all built upon the same plan and are very similar to one another, though the hindmost of the three pairs is rather the largest. The form of the proximal joint (coxa) is noteworthy; it is much flattened and can be closely apposed to the ventral surface of the thorax, being thus less prominent and less likely to impede the insect's progress in narrow crevices. The flattened shape suggests that it may be of further use as a shovel in scooping back *débris* from beneath the body when the cockroach is employed in scratching holes and enlarging small crevices. Similar enlarged and flattened coxal joints are found in the legs of many of the Fossorial Hymenoptera, *e.g. Pompilidæ*, which dig holes in the soil, and provision the nests thus formed with spiders, flies, and other insects on which the larvæ feed. I have frequently observed these Pompilids using their coxæ in the manner just mentioned, and from time to time backing out of the burrow with a heap of loose sand dragged along by each pair of coxæ, and kicked wide of the entrance by the whole leg so soon as the insect is well clear of the tunnel.

The third and fourth joints of the legs (femur and tibia), especially the latter, have upon their surfaces numerous stiff projecting bristles whose use is at once evident if dust or any fine powder is thrown over the insect. The foreign matter is without delay removed; the antennæ are drawn repeatedly through the closed mandibles and wiped clean, and then the legs are called into play and used as brushes or combs to scrape every part of the body, the tibial bristles being chiefly utilised in this work of cleansing. From time to time the legs themselves are relieved of their load of dirt either by the two members of a pair being rubbed together, or by one leg being drawn between two others closely approximated, or, again, by the action of the mandibles. It will be noticed that all the bristles point towards the apex of the leg and so foreign matter caught by them is inevitably driven towards the claws and is thus cleared away by any friction of the leg.

The cleansing contrivances of insects are variously situated and of varying degrees of perfection. Common house-flies are often seen diligently cleaning their wings and bodies by a series of brushing movements of the legs, followed by rubbing the legs together and thus casting away the collected dirt. In their case there is a more definitely specialised brush; the bristles on the leg are far finer and more closely set, and, in the absence of suitable mandibles, the legs are obliged to free each other of the dust collected. In some insects there are situated upon the front legs very perfect combs specially adapted for cleaning the antennæ; for example, many beetles, belonging to the section Geodephaga, have a deep notch on the inner side of the front tibiæ, the margins of this notch are furnished with comb-teeth which effectually clean the antenna when it is drawn through the notch. It is however among the sting-bearing (Aculeate) Hymenoptera, such as ants, wasps and bees, that the highest perfection is reached in this particular (Figs. 15 and 16). At the apex of the tibiæ of these insects are

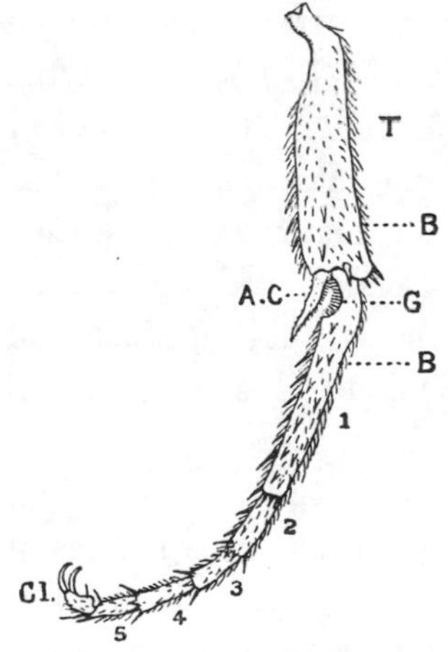

Fig. 15. Portion of front leg of *Vespa germanica* ♀. Magnified.
A.C. Antenna-comb. B B. Lines marking portion represented more highly magnified in Fig. 16. Cl. Claws on last joint of tarsus. G. Semicircular notch in which antenna is placed. T. Tibia 1, 2, 3, 4, 5, 1st to 5th joints of tarsus.

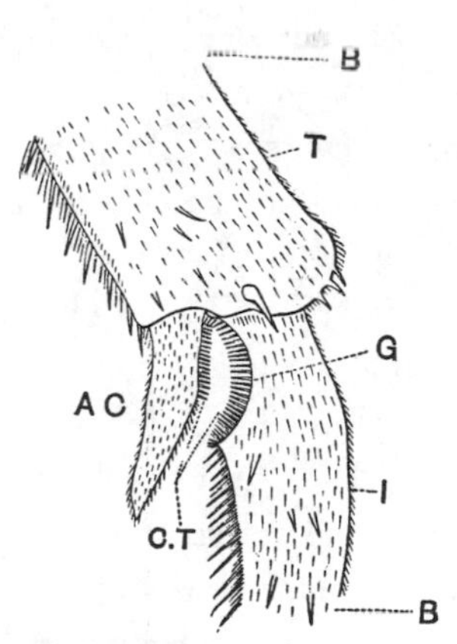

Fig. 16. The part of Fig. 15 between the lines B B more highly magnified.
C.T. Comb-teeth: other letters and figures as in Fig. 15. Note the direction in which all the bristles, both large and small, slope.

placed two stiff spines; those of the middle and of the posterior pair of legs are in no way peculiar and probably merely play a part in the ordinary cleaning of the surface of the body and legs themselves, but on the anterior tibiæ one of these spines, *A C* (Fig. 16), is much flattened and its inner margin drawn out into a closely set series of teeth (*C T*, Fig. 16), thus constituting a very beautiful comb: this structure hangs over a deeply cut semicircular notch (*G*), whose concavity is furnished with similar teeth, placed at the base (*i.e.* upper end) of the first joint of the tarsus, and almost in the angle between the tibia and tarsus. If an ant or bee or any other Aculeate Hymenopteron be watched when cleaning itself it will be seen that the foreleg is from time to time hooked over the antenna and passed along it. At such times the antenna is placed in the semicircular notch of the tarsus and at the same time the tibial comb-spine is placed against the antenna on the opposite side. The antenna is thus combed and cleaned on every side of its cylindrical surface simultaneously. It is interesting to find these insects, which are undoubtedly the highest of the Class in intellectual power, and which are almost unique in laying up stores of food for their developing young (and, in some of the best known, tending them with infinite care and foresight), should thus be provided not only with a poisonous sting for offence and defence but also with an elaborate contrivance for keeping in an efficient condition the all-important sensory organs—the antennæ.

There can be no doubt that the pollen brushes which are so characteristic of the legs of many bees are but specialised cleaning hairs. Here they are used as in other insects for removing foreign substances from the surface of the body, but in this case the extraneous matter is retained upon the legs, and the hairs are not only much longer but are beautifully branched, and thus entangle and hold the pollen more securely. The subsequent removal of the pollen by the mandibles is but a repetition of the *modus operandi* already mentioned in speaking of the cockroach. Nor is it difficult to imagine how the habits and structural modifications of the bees may have been evolved from those of more lowly but still punctiliously clean ancestors. Other insects put the bristles to different uses. For instance, among the Diptera or two-winged flies we find the bristles of the male *Platychirus* modified upon the tibiæ and tarsi so as to form adhesive pads employed in holding the female. In *Tachydromia arrogans* they assist in grasping the prey, while in *Dolichopus* they form so fine a down that the air is entangled in them and the insect enabled to glide unwetted upon the surface of water.

To return to the legs of the cockroach. Beneath the joints of the tarsus are soft white patches, resembling velvet, which prevent slipping and give the animal a foothold when running on smooth surfaces; the terminal joint carries two sharply curved claws which render locomotion possible on rough vertical surfaces, and between the claws is a pad, the pulvillus. Many insects, such as flies, are able to walk with ease on perfectly smooth vertical or even inverted surfaces. It has been found that the soles of the feet (pulvilli) of such insects emit minute drops of a gluey substance, which quickly hardens in the air, and that it is by this substance that they retain their hold. The feet do not act as sucking discs, like the familiar india-rubber contrivance for fastening a candle to the window of a railway carriage, for it has been shown by experiment that their adhesive power is in no way impaired by the removal of atmospheric pressure.

When a cockroach runs (it seldom walks) the legs are moved in a definite order, and the functions of the three pairs differ. The middle leg of one side is moved forward at the same moment as the anterior and posterior legs of the other, this constituting one step. Three legs at least are therefore always in contact with the ground, the tarsus being the part that bears the weight: the animal adopting a "three-point system" of support is in fact a tripod, when running. The legs are so placed that the front leg is extended when it advances, but the hind leg on the contrary is contracted, to bring about movement in the same direction, the femur and tibia of the latter being brought close together. Accordingly, when the whole body is moved forward by these limbs, the front leg pulls and becomes bent up while the hind leg pushes and becomes extended (this last feature is seen in an exaggerated condition in the nearly allied relatives of the cockroach, the grasshoppers and locusts). The middle leg of the other side serves mainly as a support for the body.

Food and digestion. Cockroaches will eat almost anything—all that is food for man, whether of an animal or vegetable kind, paper, leather, refuse of every kind and description, and even the dead bodies of their fellows. They are also said to devour bed-bugs and are requested as a favour from sailors by native tribes, who are greatly troubled with those parasites. The strong teeth at the tips of the mandibles and the roughened grinding surfaces near the base of their inner edges fit more or less closely into one another right and left, and appear not only to bite off pieces of food but also to grind them into fragments sufficiently small to pass along the narrow gullet into the large crop. Prior to being swallowed, the food is moistened by the faintly alkaline secretion of the salivary glands, which not only lubricates and softens but is also an actively digestive fluid capable of converting starch into sugar. Further digestion is brought about in the crop into which there passes forward the feebly acid secretion of the seven or eight hepatic ("liver") tubes which are connected with the chylific stomach at its anterior end close behind

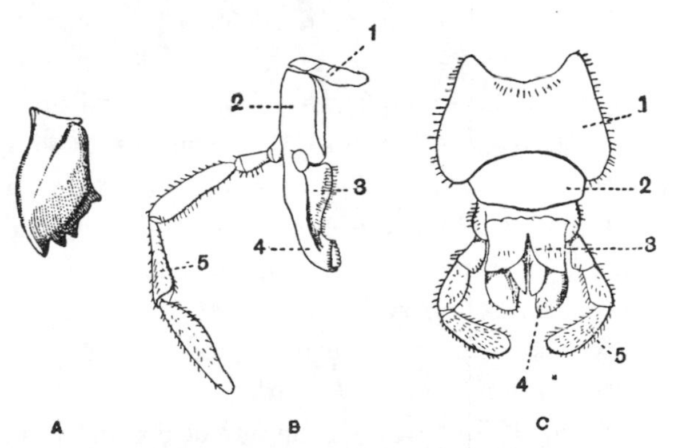

Fig. 17. Mouth-appendages of *Periplaneta*. Magnified.

A. Mandible. B. 1st maxilla. 1. Cardo. 2. Stipes. 3. Lacinia. 4. Galea. 5. Palp. C. Right and left 2nd maxillæ fused to form the labium. 1. Submentum. 2. Mentum. 3. Ligula, corresponding to the lacinia. 4. Paraglossa, corresponding to the galea. 5. Palp.

the gizzard. This secretion emulsifies fats and oils, renders proteids soluble, and completes the conversion of starch into sugar. Thus all ordinary food substances are made capable of diffusing through the walls of the crop and of the chylific stomach into the blood. Examination of the fæces gives reason for thinking that at least some cellulose undergoes digestion, but I am not aware of any exact observations on this point. In the gizzard a certain subsidiary amount of grinding is effected by the six hard longitudinal ridges projecting from the inner face, but the chief use of this organ is to act as a strainer and allow only very fine particles to penetrate the sieve formed by its fine interlacing hairs. By the time the food reaches the rectum, all digestible substances have been rendered soluble and absorbed into the blood; the fæcal residue is moulded into pellets in this last section and expelled *per anum.* The six longitudinal ridges which project into the rectal cavity probably assist the expulsion by giving the walls a firmer grip upon the contents and also give the slight screwing movement observable in the fæces as they are forced out. Similar screw-like twists are to be seen as the solid and liquid ejecta of many other animals leave the body. The muscular effort required to effect any such expulsion is appreciably lessened in intensity by calling into use the mechanical principle of the screw.

Excretion. It is convenient here to deal with the system for excretion of nitrogenous waste material inasmuch as the Malpighian tubules, which are the organs concerned, are attached to and open into the posterior end of the chylific stomach near its junction with the intestine.

These tubules are intricately entwined among the other abdominal viscera, and extract from the blood in which they are bathed various nitrogenous compounds, uric acid being that most generally known.

Respiration. The air enters and leaves the body not by the mouth but by the paired openings (stigmata, spiracles) situated upon the sides of the body. By means

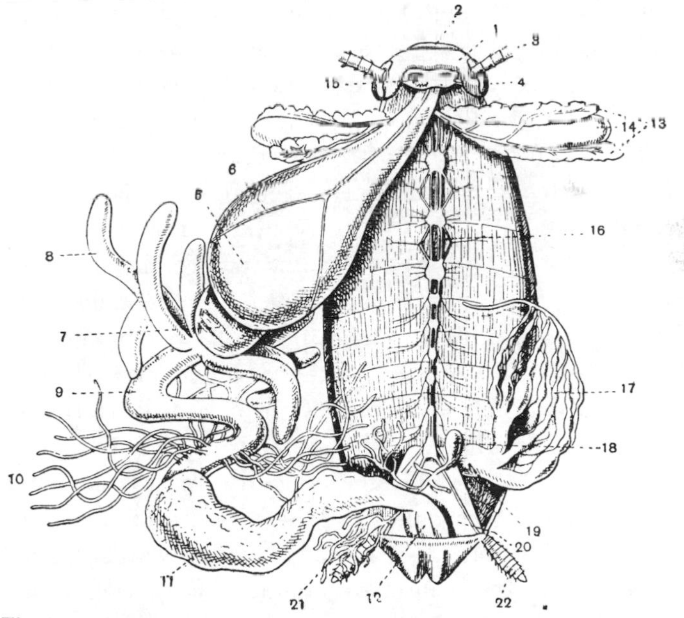

Fig. 18. A Female Cockroach, *Periplaneta*, with the dorsal exoskeleton removed, dissected to show the viscera. Magnified about 2.

1. Head. 2. Labrum. 3. Antenna, cut short. 4. Eye. 5. Crop. 6. Nervous system of crop. 7. Gizzard. 8. Hepatic cæca. 9. Mid-gut or mesenteron. 10. Malpighian tubules. 11. Colon. 12. Rectum. 13. Salivary glands. 14. Salivary receptacle. 15. Brain. 16. Ventral nerve-cord with ganglia. 17. Ovary. 18. Spermatheca. 19. Oviduct. 20. Genital pouch in which the egg-cocoon is found. 21. Colleterial glands. 22. Anal cercus.

of the tracheal tubes, of which these are the external apertures, the air is distributed to all parts of the body direct. The main tubes, previous to branching, in each segment are connected by longitudinal trunks with those of the other segments, so that in the event of one spiracle becoming accidentally blocked the segment normally supplied through that spiracle is able to obtain fresh supplies of air through the neighbouring apertures. The animal is able voluntarily to close the spiracles. In order that inspiration and expiration may be effected certain more or less rhythmic movements of the abdomen are executed. Of these the most conspicuous is the flattening of the abdomen, the dorsal and ventral surfaces being brought closer together by the contraction of the dorso-ventral muscles, which are able to reduce the depth of the body by as much as one-eighth; at the same time a lateral compression also takes place. By this mechanism the tracheal tubes are squeezed and the air driven out at the spiracles. When the muscles relax the skeletal parts return to their original positions, the tracheal tubes widen out again in virtue of their elasticity due to the spiral thickening of their walls and, the pressure of air within being thus less than that of the atmosphere without, fresh air rushes in. In the ultimate ramifications of the tracheal tubes the air is renewed by diffusion rather than by a direct inrush, in much the same way as the "tidal air" inspired by us does not by any means fill the lungs with entirely fresh air but mixes with and diffuses into that which remains in the lungs and which cannot be driven out by any effort on our part.

All, except some aquatic insects, exhibit dorso-ventral flattening of the abdomen in expiration, but in some species other movements are equally, if not more, evident. Dragon-flies for example diminish the transverse horizontal dimension of the abdomen strongly, while the Aculeate Hymenoptera alternately shorten and lengthen the body by telescoping the segments more or less.

Besides the special respiratory movements, the normal contraction of the muscles of the body and limbs in locomotion must assist in the movement of the air within the body; every muscular contraction causes a pressure upon the tracheal tubes in that region, and forces the air along the tube, perhaps both outwards and inwards, and to some extent assists in the respiratory interchanges.

The true respiratory movements are entirely reflex actions, persisting in the decapitated insect and even in the detached abdomen, as may often be noticed in a wasp bisected on a dinner plate, and in other species where the nervous system has not undergone concentration. In such cases the movements are hastened or retarded by the same causes (heat, cold, and so on) as hasten or retard them in the intact insect.

Circulation. The direct supply of air to the tissues of the body relieves the blood from the duty of conveying oxygen and carbon dioxide, and thus the relative insignificance of the blood system receives an explanation. The blood is colourless and is kept flowing through the irregular channels and spaces among the tissues by the contractions of the dorsally situated, chambered, tubular heart, which receives the blood through a pair of valved openings in each segment, and forces it forward towards the head, all backward flow within the heart being prevented by the funnel-like communication of chamber with chamber. The rate of heart-beat varies, but is on an average from 70—80 beats per minute. Around the heart is a space partitioned off from the rest of the body spaces (blood cavities), excepting at certain holes through which the blood streams when the pericardial space is enlarged by the pull of the paired, fanshaped alary muscles in each segment: the handle of each "fan" is attached to the side of the dorsal covering of the body while the expanded part is spread out below the heart in the partition membrane. When the heart is relaxed blood flows into it from the pericardial space through the valved openings. Thus by the alternate contractions of the alary muscles and of the wall of the heart the circulation of the blood is maintained. Another space with pulsating walls of very similar structure covers the ventral nerve cord and urges the blood from the head towards the tail by waves of contraction, which run from before backwards.

Nervous System and Special Senses. The cerebral ganglia (brain) are the centres of all voluntary action, and exercise, as in the crayfish, a general controlling influence over the activities of the ganglia of the ventral chain, which are, to a very great extent, reflex centres. Reference has already been made to their action in connexion with the respiratory movements; they are also responsible for the protrusion of the sting in decapitated wasps and their allies, and again for the phenomenon of "shamming dead" that may be witnessed in many insects.

The common Currant Moth (*Abraxas grossulariata*) is a notable malingerer, and will feign death even after decapitation. In 1895 I decapitated a specimen that was feigning death, and was able to keep it in a responsive condition for two days, during which period any stimulus, such as touching or pinching, induced a repetition of the death-feigning: a strong stimulus was followed by a prolonged, and a weak one by a more brief quiescence. Eventually a weak fluttering movement set in for some hours, followed by death.

In this case it is clear that the sensory stimulus received by the surface of the body caused inhibitory impulses to arise reflexly in the ganglia of the ventral nerve chain and prevent all movement of the locomotor muscles. Areas of the wings which had become denuded of scales showed a marked diminution in sensibility, thereby indicating that the scales are tactile organs. The scales of Lepidoptera are merely highly modified representatives of the hairs or bristles that are frequent on the limbs and bodies of other insects where they function as organs of touch.

This sense is in the cockroach and in all insects very highly developed in the antennæ and in the palps of the 1st maxillæ and labium. The antennæ and the palps are kept constantly flickering over the surface on which the cockroach is travelling. From time to time the former are waved in the air, in a manner suggestive of their possessing also the sense of smell, as indeed is almost certainly the case, some 39,000 sensory rod-like nerve endings being situated in each antenna of the male. The palps are also in all probability organs of touch and of smell rather than of taste, this latter sense being apparently possessed by the margins and walls of the mouth itself. At the posterior end of the body the two cerci are extremely sensitive to touch, and may perhaps act as sentinels in the rear and possibly also have special tactile functions during coitus.

To what extent the compound eyes are capable of forming a distinct image of surrounding objects we cannot say, but it is evident that they are keenly sensitive to differences of light and shade from the speed with which a cockroach makes for dark corners and crevices when disturbed. That some insects obtain clearly defined visual impressions seems indisputable. I once observed a Brimstone butterfly visiting flowers of the Dog Violet scattered along a bank, and picking out these flowers to the exclusion of all others with great precision, not approaching even other blue flowers that were present.

No auditory organ has been discovered in the cockroach, nor is it probable that they are capable of hearing, and in this connexion it is interesting to note that they are likewise incapable of emitting sound. There is a very great diversity among insects in regard to the position and structure of auditory organs and of contrivances for the production of sound.

Space does not permit us to enter fully into this interesting subject, and it must suffice to deal only with these structures in the grasshoppers, locusts and crickets. The chirrup of the grasshopper is produced, as may easily be observed on any bright day in summer-time, by the movement of the hind-legs: the inner face of the femora of these legs bears, on a ridge projecting inwards, a series of knobs, which are modified hairs. When the femora are moved up and down these knobs strike against a sharp-edged vein which stands out prominently from the outer face of the upper wings, and thus the well-known note is emitted. The auditory organ which enables neighbouring grasshoppers to hear the call is situated on the dorsal side of the first abdominal segment, just above the articulation of the hind leg: it consists of a broad slit opening into a cavity with the auditory membrane below.

Locusts produce their note in a different manner: the left fore-wing has on its inner surface a file which is worked to and fro over a sharp edge on a prominent part of the inner margin of the right fore-wing. The auditory organ, too, is not situated as in grasshoppers, but lies in the upper part of the tibiæ of the front legs. In the crickets each fore-wing has on its under surface a stridulating file, the right as a rule overlapping the left: the auditory organs consist of a pair of cavities on each front leg, a larger on the outer and a smaller on the inner face. Some authorities state that there is yet another pair of auditory organs on the second abdominal segment.

Reproduction and growth. The breeding season of the cockroach is during the summer months. The spermatozoa of the male are passed into the spermatheca of the female in the course of sexual union, the male obtaining a secure hold upon the female by means of a series of strong hooks and plates surrounding the genital aperture. The ventral portion of the seventh segment of the female is produced backward into a large boat-shaped process, which forms the actual extremity of the body, the corresponding parts of the eighth and ninth segments being telescoped inwards so as to lie dorsal and anterior to that of the seventh, and similarly the ninth to the eighth. The eggs when mature pass out of the opening of the common oviduct by an aperture in the ventral part of the eighth segment into the cavity formed by the seventh; at the same time spermatozoa are poured from the spermathecal opening, which is in the ninth segment and almost directly above the oviducal aperture. Sixteen eggs, one from each of the sixteen ovarian tubules, mature and are fertilised at about the same time, and pass into the genital pouch formed by the seventh segment: the walls of this pouch are lined by the secretion of the "colleterial glands," which hardens in that position and forms as it were a hollow cast of the pouch itself. Into this cast—the future egg-capsule—the eggs are passed one by one, until the capsule contains sixteen eggs, arranged in two alternating rows of eight each; the inner end of the capsule is then closed, and the whole is ready for deposition. It is now a dark, mahogany coloured, horny structure of oblong shape, with rounded ends, and a toothed longitudinal ridge along its upper surface; it is about half an inch long, and a quarter of an inch wide and high. Females are often seen carrying the egg-capsule half protruding from the genital pouch, and are presumably seeking a favourable place and opportunity for depositing it. When this is accomplished they take no further interest in their progeny.

The eggs are upright within the capsule, the future

head of the embryo is uppermost, and the ventral surface of one row faces that of the other, the dorsal surface of each being thus against the side walls of the capsule. When the young are ready to emerge, they push their way out of the capsule at the longitudinal ridge, which is in reality the thickened margins of a slit cemented together. The cement is said to be previously softened by a fluid secreted by the embryos. After the escape of the young the slit is closed up by the elasticity of the walls. The young cockroaches are almost colourless at first, excepting the eyes, which are black.

The first moulting of the skin takes place immediately after hatching, the second about a month later, the third at the end of the first year; subsequent moults occur once a year only in the warm weather. The split for the escape from the old integument extends along the middle of the back of the thoracic segments. After each moult the skin is soft and pale, but soon hardens and gets darker in colour. Young cockroaches are always of a lighter tint than adults. Maturity is not reached until the fourth year, according to some observers, though in *Bl. germanica* the immature stages are passed through more rapidly; nevertheless it is difficult, in the face of the apparent rapid production of new generations of cockroaches, to believe that so long a period is the invariable rule with *Bl. orientalis*. The young males are at first, like the females, wingless, and only acquire their wings after several moults, but they may at all times be distinguished from the females by the possession of a short pair of styles projecting backwards from the ninth sternum just below the cerci.

Foes and Parasites. In spite of their evil odour a fair number of animals are known either on occasion or habitually to feed upon cockroaches; hedgehogs of course are proverbially successful in clearing infested premises of these pests, but rats, cats, polecats, some birds, frogs and wasps have been known to devour them, and some few Fossorial Hymenoptera lay up stores of them for the benefit of their larvæ.

A considerable number of parasites have been found in various parts of the body. In the rectum and intestine six species of Protozoa, viz. *Endamœba blattæ*, *Gregarina* (*Clepsidrina*) *blattarum*, *Nyctotherus intestinalis*, and three Infusorians, *Plagistoma blattarum*, *Lophomonas blattarum*, and *L. striata*. From the intestine of *Bl. americana* Schuster[1] has described yet another Infusorian which he names *Lophomonas sulcata*. In addition to bacteria and *Hygrocrocis intestinalis*, an alga, several Nematode worms of the genus *Oxyuris* occur, *Gordius* also is found, while encysted in the fat-body is found *Filaria rhytipleuritis*, which becomes sexually mature in the rat. The other most notable parasites are a beetle (*Symbius blattarum*), a hymenopteran (*Evania*), and an acarid that frequents the male reproductive organs.

The forms allied to the Cockroach—British Orthoptera[2].

The Cockroach belongs to the order of insects known as ORTHOPTERA: the two sections of the order are distinguished as Cursorial and Saltatorial, according as they walk or leap.

[1] *Proc. Zool. Soc.* Lond. 1898.

[2] M. Burr, *British Orthoptera*, Huddersfield, 1897.

The Cursorial forms have the hind-legs adapted for running, and the wing cases (fore-wings), when present in the immature stages, lie flat. We have two British Cursorial families—the Earwigs (Forficularia) and the Cockroaches (Blattodea). The Earwigs have their flight-wings folded when at rest not only fanwise but also transversely in a complicated manner; the cerci at the hinder end of the body are converted into nippers and the tarsi of the legs have but three joints. The Cockroaches on the other hand have the wings folded only fanwise, the cerci are jointed and not modified into nippers, and the tarsi have six joints.

Eight species of earwigs have been recorded in England, but only two of these are at all common, viz. *Labia minor* and *Forficula auricularia*; of these the former is small, about 5 mm. long, and has the second joint of the tarsi cylindrical. It may often be taken on the wing on summer evenings flying over flowers. The latter, the common earwig, is 10—15 mm. long, and has the second joint of the tarsi lobed: it is distinguishable from *Forficula pubescens*, which has occurred in many places in England, but cannot be called common, by possessing perfect wings and 15 joints in the antennæ, whereas *F. pubescens* has abortive wings and only 12 joints in the antennæ. Other species which may possibly be met with near the coast and especially among the shingle are *Labidura riparia*, of a pale red colour, with well-developed wings and 27—30 joints in the antennæ; *Anisolabis maritima*, of dark brown colour without wings, and with 24 joints in the antennæ; and *A. annulipes*, black in colour, devoid of wings, antennæ with 16 joints only, and yellow at the base, then black, the 13th and 14th joints pale, 15th and 16th dark; the legs also have dark rings round them. The two remaining species are *Apterygida albipennis* and *A. arachidis*. They are probably not truly indigenous and have only been found very rarely. The genus can be distinguished by the nippers of the male being separated at the base and not flattened.

Common earwigs live mostly on decaying vegetable matter but also eat petals of flowers, leaves, ripe fruit and similar substances; they come in swarms to "sugar" placed on trees for the purpose of catching moths. The female watches over her eggs and offspring for some time; the eggs are deposited underground and hatched in early spring, and full growth is attained in about three months' time. The nippers of the female are weak and nearly straight, those of the male are sharply curved and far stronger; the use of these organs is not fully understood, though it is known and can, with some patience, be observed that they are used in neatly disposing the flight-wings beneath the wing-covers. They are undoubtedly used defensively if an earwig be picked up, but the pinch given is so feeble that it is difficult to imagine that they are of any real value in this respect; I have seen an earwig use its nippers vigorously but to no purpose in vainly resisting an attacking wasp. It seems probable that they may play an important part in sexual union. An analogous, though very differently situated apparatus, is found in the male stag-beetle; here the mandibles are modified to formidable looking pincers which however are not capable of inflicting a severe wound on the finger, but are used, as I have witnessed, to hold the female during sexual union; they also gave a terrifying aspect to the insect sufficient to deter some enemies.

Of the ten species of Blattodea that have occurred in Britain only three are truly indigenous; the remainder, including the cockroach (*Bl. orientalis*), having been imported, probably by shipping. The three indigenous species all belong to the genus *Ectobia*; they are much smaller than the cockroach and do not frequent our dwellings. (1) *E. lapponica* has the dorsal surface of the first thoracic segment (pronotum) darkly coloured except along the edges, where it is pale : the insect is from 8 to 11 mm. long, and lives in shrubs, under moss and dead leaves. The two other species, which both have the pronotum pale, are (2) *E. panzeri*, 6—8 mm. long, greyish in colour, the wing cases of the female reach only to the third segment ; found under refuse and rubbish in sandy districts and sand dunes by the sea-shore : (3) *E. livida*, 8—8·5 mm. long, pale straw colour with a reddish tint on the pronotum, wing covers reaching beyond the apex of the abdomen in both sexes, and flight-wings well developed in both.

Of the remaining genera and species most are very local, occurring in a few spots, such as a warehouse, or a dockyard, where they have been introduced. We may however mention the large *Periplaneta americana*, which attains a length of about 3 cm. and is of a bright brown colour with well-developed wings and wing-covers longer than the abdomen in both sexes : it is common in many of the houses in the Zoological Gardens and is probably known by sight to many persons.

The Saltatorial section includes all forms with hind-legs adapted for leaping: additional diagnostic characters are furnished by the presence of stridulating organs, the possession by the female of an ovipositor, and by the wing-covers in the immature forms having the lower margin turned towards the dorsal part of the insect.

The GRASSHOPPERS (*Acridiodea*) have short antennæ and 3-jointed tarsi : the ovipositor of the female is short and inconspicuous : the stridulating organs are partly on the femora of the hind-legs and partly on the wing-covers. In flight the front and hind wings are hooked together so as to move as one. There are eight genera of British Grasshoppers but only three of these are at all common, viz. *Stenobothrus*, *Gomphocerus* and *Tettix*.

Tettix is distinguished from all the other genera by the curious extension of the pronotum backward over the abdomen : we have two British species, *T. bipunctatus* with highly arched pronotum and short wings, and *T. subulatus* with nearly flat pronotum and long wings. These are both small grasshoppers, 7 to 10 mm. long, and vary greatly in colouring. They appear in the spring and early summer and hibernate in winter: the eggs are laid in spring. They frequent dry barren places among dead leaves or similar spots.

Stenobothrus, which includes 6 species distinguishable from one another by minute characters into which we cannot now enter, differs from *Gomphocerus* (3 British species) in possessing thread-like tapering antennæ, whereas in the latter genus they are club-shaped.

The LOCUSTS (*Locustodea*) and CRICKETS (*Gryllodea*) are alike in possessing long antennæ, a conspicuous ovipositor of greater or less length in the female, and stridulating organs confined to the wing-covers.

The two families differ from one another in the following respects:

The Locusts have 4-jointed tarsi, and the stridulating part of the wing-cover is smaller than the rest.

The Crickets have never more than 3 joints in the tarsi, and the stridulating part of the wing-cover is larger than the rest.

The *Locustodea* include various genera and species of "Sword-tailed-" and "Sabre-tailed-Grasshoppers," and the "Green Locust." Eight genera have been recorded in Britain; the more common are the following:

Leptophyes punctatissima, green; male 12 mm., female 16 mm. in length; wings abortive in both sexes, 1st and 2nd joints of tarsi smooth and not grooved at the sides. The last character distinguishes this form from all other common British Locustodea. The species is found on trees and shrubs in summer and autumn: the eggs are laid in crevices in the bark, and hatch in May.

Meconema varium, pale green with a yellow line along the back; 11—15 mm. in length, aperture of auditory organ on anterior tibiæ wide open. Male devoid of stridulating organs. Occurs on trees, especially lime and oak, in late summer and autumn.

Xiphidium dorsale, bright green, with reddish tint dorsally, body 12—15 mm. in length; aperture of auditory organ reduced to a narrow cleft; stridulating organs present in male; anterior tibiæ smooth at the sides and devoid of apical spine on outer margin. Occurs but locally in damp spots, marshes and river banks in late summer and autumn.

Locusta viridissima, dark green, with reddish markings; male 20—23 mm., female 32—35 mm. in length of body; aperture of auditory organ reduced to a cleft, stridulating organs present in male, anterior tibiæ grooved at the sides and spined on the outer margin at the apex; ventral portion of 1st thoracic segment (prosternum) carries two spines. Occurs in coarse herbage, nettle beds, etc. in summer; is fairly common along the south coast. I have also taken it in Surrey (Godalming) and Oxfordshire.

Thamnotrizon cinereus, brown, with black markings; male 13—15 mm., female 15—18 mm. in length of body: auditory apertures as in the two last, anterior tibiæ as in *L. viridissima* but with 3 spines, prosternum not armed with spines. Wing-covers and wings scale-shaped. Occurs in bushes in autumn and often chirrups in the evening and on into the night.

Platycleis grisea and *P. brachyptera* resemble *Thamnotrizon* in all structural points, but the wing-covers and wings are never scale-shaped. In *P. grisea* the wings and wing-covers are well developed; colour greyish brown, length 17—22 mm. Found in dry barren spots especially in chalky districts; season autumn. *P. brachyptera* has very short wing-covers and wings, is dark brown with green and black markings; length of body 12—16 mm., occurs in late summer and autumn on heaths, commons, clearings in woods, etc.

The *Gryllodea* are represented by 4 species only:

Nemobius sylvestris, the Wood Cricket; chestnut brown with lighter markings; length of body of male 2 mm., of female 2·5 mm., anterior feet adapted for walking; spines of posterior tibiæ slender and movable; posterior tarsi not grooved nor serrated, wing-covers short. Occurs in New Forest; is adult in July: among dead leaves, etc. on dry banks.

Gryllus campestris, the Field Cricket; nearly black; length of body 20—26 mm., anterior feet adapted for walking; spines of posterior tibiæ stout and immovable ; posterior tarsi grooved above and serrated at the sides. Wing-covers well developed and with a yellow spot at the base. Flight-wings shorter than the covers. In hot sandy places through the summer.

Gryllus domesticus, the House Cricket, resembles the last species in all chief structural points, but is paler in colour and has brownish wing-covers, the flight-wings are fully developed and longer than the abdomen. The length of the body is only 16—20 mm. This species frequents the neighbourhood of fireplaces and ovens—the Cricket on the Hearth—and may be found in all stages at all seasons of the year. It is far less common than formerly and is perhaps being ousted by the Cockroach.

Gryllotalpa gryllotalpa, the Mole Cricket ; reddish brown with darker markings, length of body 35—50 mm. ; anterior feet adapted for digging ; the female has no projecting ovipositor. Lives in holes in the ground ; is found chiefly in the southern counties but is not common ; feeds on roots and also on animal substances. The chirrup is to be heard on warm evenings in the spring and early summer.

CHAPTER IV.

DRAGONFLIES.

THE Cockroach and other Orthoptera mentioned in the last chapter exhibit in the course of their development none of the sudden and abrupt changes of form which constitute "metamorphosis." From birth they resemble the adult in all but a few features, and these are acquired gradually at successive moults and not merely at one or two. Up to a certain point the Dragonflies exhibit similar phenomena, but eventually there comes a well-marked change both in appearance and habitat when the insect passes from the "nymph" (pupa) stage to that of the adult (imago). There is however no period of prolonged quiescence and immobility such as is familiar in the chrysalis of a Butterfly or Moth, and they are therefore described as having only a "partial metamorphosis."

The entire group of Dragonflies is known as the ODONATA and is best regarded as a distinct Order of Insects. In England we have about 40 species. The great majority seldom travel far from water, but some of the larger species with strong powers of flight at times wander great distances and are even known to migrate in great swarms, though the object of migration is entirely unknown.

The main structure and habits of all are very similar. Possessed of both simple and large compound eyes (whose facets in some cases are larger upon the upper than under side, the former perhaps for remote and the latter for near vision) they have keen powers of sight and rely upon this sense for the capture of their prey, the antennæ being (consequently?) reduced. The wings are relatively large and strong; both pairs are used in flight, but there is no apparatus for linking the fore and hind wing of the same side together. Nevertheless, in many species, the flight is remarkably bold and swift and enables these insect-hawks to catch with ease the flies and other insects on which they feed. The front and hind wings are free from each other[1], and are provided with their respective elevator and depressor muscles. But they are perhaps brought into working connection with each other by means of a long lever present among the chitinous pieces in the body-wall at the base of the wings.

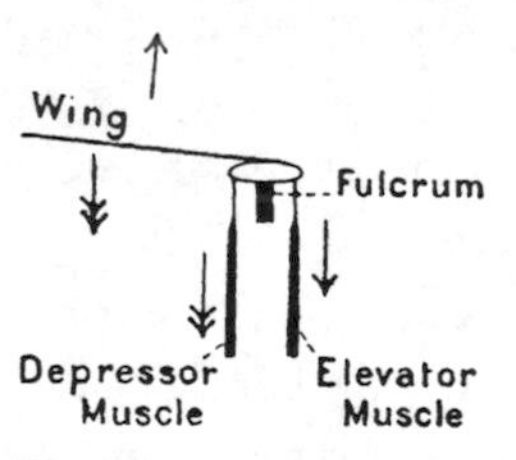

Diagram to explain action of wing of Dragonfly.

Von Lendenfeld has described the arrangement of the muscles[2], ligaments and complicated chitinous joints concerned in the movement of the wings. If reduced to a simple plan the mechanism appears to be that of a lever

[1] Chabrier, "Vol des Insectes," *Mem. d. Mus.*, 1822.

[2] *Akad. d. Wiss.* Vienna, 1881, Hft. I.

with the fulcrum very close to one end. The elevator muscles are attached to the short arm of the lever—which in the upward movement of the wing is one of the first order. The depressor muscles are attached to the lever close to the fulcrum but upon the far side of it, that is to say upon the same side as the wing expansion, which is the weight to be moved. The lever is now one of the third order.

The long abdomen, which has perhaps gained for them the false reputation of being able to sting—"Horse-stingers" is one of their popular names—is brought into play in flight, enabling the creatures to turn and twist, and also in a remarkable way in the act of mating. The legs are of moderate length and furnished with a double series of long bristles; they are rarely employed for walking, but during flight are all turned forward and meet at about the level of the mouth; in this position they form, by the aid of their bristles, a sort of basket beneath the mouth for the reception and retention of flies caught in mid-air. It is difficult to determine whether the prey is seized by the jaws or by the legs; my own opinion is that the jaws are the weapons of prehension and that the legs hold the prey thus seized while its wings and less nutritious parts are cut off by the jaws and allowed to fall to the ground. I have frequently observed Dragonflies capturing their food, and a distinct "snap," such as the jaws might easily cause as they come together, is audible at the instant of capture, and in a second or two, wings and other parts of the victim are seen floating down through the air. On the other hand, if an insect is offered to a

Dragonfly held captive by its wings, it is received by the legs and brought forward to the mouth.

The acts of mating and of fertilisation are accomplished during flight or in part during rest on reeds or other herbage. The reproductive aperture of the male is situated on the ventral surface of the ninth abdominal segment. But it is not from this aperture direct that the spermatozoa are received by the female, for prior to sexual congress the male bends his abdomen downwards and forwards so as to bring the ninth segment in contact with the second which has in it a small sac into which the spermatozoa are passed. Within this sac is a penis of complicated structure; its exact form differs greatly in the various genera and species[1]. The female is then sought by the male and grasped from above round the neck by the clasping appendages (in some cases, *e.g. Lestes*, highly complex structures) situated at the posterior extremity of the body. In this phase the two insects are placed "tandem-wise," the tail of the male just overlapping the head of the female. The female now bends her abdomen round underneath so as to apply her ninth abdominal segment, which carries the opening of the oviduct, to the sperm-sac of the male in his second abdominal segment, and thus fertilisation is brought about. In some species the male retains his hold upon the female while the eggs are being deposited, even to the extent of entering the water with her; in others he releases her. The eggs are deposited in various ways. Some species drop

[1] Miss M. F. Goddard, *Proc. Am. Philos. Soc.* Philadelphia, xxxv. 1896.

them singly at random in the water, just touching the surface as they do so, others alight upon water-weeds and deposit their eggs among them, or within them by incisions, pushing the abdomen a little way into the water. Others again lie flat with wings expanded upon the surface and the abdomen hanging down into the water, while yet others do not hesitate to walk down the stems of reeds or grasses completely below the surface of the water and deposit the eggs on or in the mud at the bottom. Occasionally the eggs are laid in the mud of partly dried up ponds and ditches. The eggs are small, mostly less than 1 mm. in length, and elliptical in outline; in some the major axis is four or five times as great as the minor, in others not twice. They are not unfrequently infested by a minute hymenopterous parasite, *Anagrus incarnatus*.

In about three or four weeks after oviposition young larvæ emerge from the egg-shell. They are at first almost transparent and swim about by means of their relatively long legs, and subsequently they become opaque and move about chiefly by crawling. The skin is cast off several times. The wings, of which at hatching there was no trace, appear first as minute processes, and increase in size with each successive moult. By the time the larva is ready to pass into the imago the wings reach about half-way down the abdomen; even then however they are quite insignificant in comparison with those of the perfect insect.

The food during all stages of life consists of living animals; in adult life it is obtained by bold and vigorous flight in the air, in the earlier stages by stealthy and imperceptible approach. The larvæ of all species are most difficult to detect; some closely resemble the mud in and on which they live, others are like black water-logged bits of stick, others living among green water-weeds are so coloured as to be practically invisible amidst their natural surroundings. Thus they not only escape the notice of their foes but are able unperceived to approach their prey, which consists of soft-bodied aquatic animals, such as worms, insect larvæ, and even tadpoles and small fish. On occasion however they can advance with some speed either by forcible expulsion of a jet of water from the hinder end of the body, or in some species by swimming with the aid of three blades projecting from the posterior extremity.

The prey is seized by means of the labium (united second maxillæ) which is of peculiar construction and is known as the "mask." This contrivance lies under the head and anterior part of the thorax; it is attached, as usual with the labium of insects, below the mouth. When at rest its proximal portion is directed backward and united to the distal portion by an elbow-joint hinge from which the latter projects forward, lying ventral to the proximal portion. Thus the attached base and the free tip are close together at the mouth itself. At the free tip is a pair of sharp movable beaks (the modified labial palps) known as the forceps. When the larva arrives within striking distance of its prey, the "mask" is suddenly shot forward, the elbow-joint straightening out. The forceps seize the animal which is then by a quick return of the "mask" dragged back to the mandibles and first maxillæ, by which it is speedily cut in pieces. The action of the "mask" may be well imitated with the human arm:—keeping the upper arm against the side, place the palm of the hand against the shoulder, the whole arm is now in the position of the mask at rest; the upper arm represents the proximal part, the elbow the hinge, the fore-arm the distal part: the forceps may be represented by separating the third and fourth digits; now extend the arm straight above the head, bring third and fourth digits in contact, and return the arm to its first position.

Some of the most interesting studies in natural history are to be found in the means adopted by aquatic insects for securing the oxygen necessary for respiration. Tracheal tubes are essentially air-breathing contrivances; their lateral apertures, the spiracles, constitute efficient and convenient points for entry and exit of air in terrestrial insects, but not so in those which have betaken themselves beneath the surface of the water. Hence in these latter we find many beautiful modifications and ingenious devices for maintaining respiration.

The larvæ of all Dragonflies have two pairs of spiracles upon the thorax; in some (*Libellulidæ*) the pair between the first and second segments of the thorax can be seen on the dorsal surface with the unaided eye. It is however not until the later phases of larval life that these openings become functional; indeed in the earlier stages they are not really open but air is obtained by other means. In the larvæ of the two families, the *Libellulidæ* and *Æschnidæ,* which include our larger and strong-flying species, and whose members may be at once known by the fact that the *imagines* spread their wings flat when at rest, respiration is carried on by the hinder portion of the intestine, the rectum, which alternately receives and expels water through the anus. This latter is surrounded by five valves, the three larger of which can be brought together or widely separated so as to guard or expose the terminal opening of the rectum. Six longitudinal bands of muscle run in the rectal wall, each carrying several thousand transverse folds which very largely increase the superficial area and also receive fine branches from the tracheal tubes. The intervals between the six bands are occupied by very thin and flexible cuticle. The whole arrangement permits of considerable distension so that a quantity of water can be drawn into the rectum. The oxygen dissolved in the water passes through the walls of the rectum into the tracheal branches and so is distributed over the body. It is by the forcible expulsion of this water of respiration that the larvæ can, if they wish, propel themselves forward through the water at a moderate speed.

The *Agrionidæ* have, in place of the five valves, three thin blade-like lamellæ of considerable length; these, as already mentioned, are employed in swimming, but they also take an important part in respiration. Each lamella possesses a close network of tracheal tubes which doubtless absorb air that is dissolved in the water and pass it on to the main tracheal branches, and so supply all the tissues of the body. Some Agrionids, and perhaps all, can also take water into the rectum for respiratory purposes. In the later stages of larval life the spiracles are certainly often functional and probably the insects can and do breathe at will either through them or by the apparatus just described. At these periods they not unfrequently come to the surface or even climb out of the water a short distance and thus expose the spiracles directly to the air. If the water is foul, and presumably deficient in oxygen, an Æschnid larva will come to the surface and take air into the rectum by protruding the extremity of the abdomen.

At the close of larval life the animal ceases feeding for a time and climbs up a reed or post or other upright object projecting from the water, or may even travel a few yards over dry land until it meets with a suitable support up which to crawl. Having gained a secure foothold by means of its claws the larva rests with its head uppermost for some hours without undergoing any visible change. The larger species when resting on rushes resemble the brown clusters of the flowers of these plants and are thus protected during the period of exposure and helplessness. The first noticeable alteration of appearance takes place in the eyes which, from being dull and more or less opaque, become bright and clear in consequence of the underlying brilliant eyes of the imago pushing against the larval covering. Shortly afterwards a crack appears along the mid-dorsal line of the thorax, and the thorax of the imago begins to protrude, extending the cleft forward on to the head and causing lateral splits in the region of the eyes. The thorax (with the, at present, diminutive wings) and head are freed first from the larval skin and then by a strong bend dorsalwards the legs are extracted.

Thus much accomplished, the imago throws itself backward, no longer retaining any hold by means of its legs, but supported solely by the still confined hinder segments of the abdomen. This becomes sharply bent over the posterior margin of the crack through which the creature has emerged. The inverted position is maintained for about half-an-hour, probably until the legs and claws have become dry and attained the desired firmness. At length, with a jerk, the animal springs up into its previous atti-

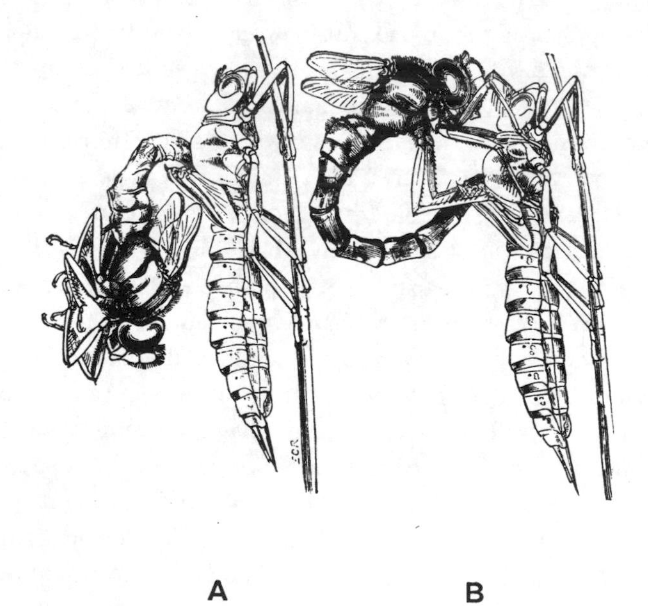

Fig. 19.

A. The anterior portion of the body of a Dragonfly, *Æschna cyanea*, freed from the larval shell. B. The tail being extricated.

tude; or at times the head and thorax are repeatedly jerked up and allowed to drop back again in rapid succession; the legs, now strong enough for the purpose, grasp the sides of the empty larval skin, the abdomen is withdrawn from its case by a strong dorsal curvature, and the imago stands free upon the larval husk.

The freshly emerged imago is colourless and almost

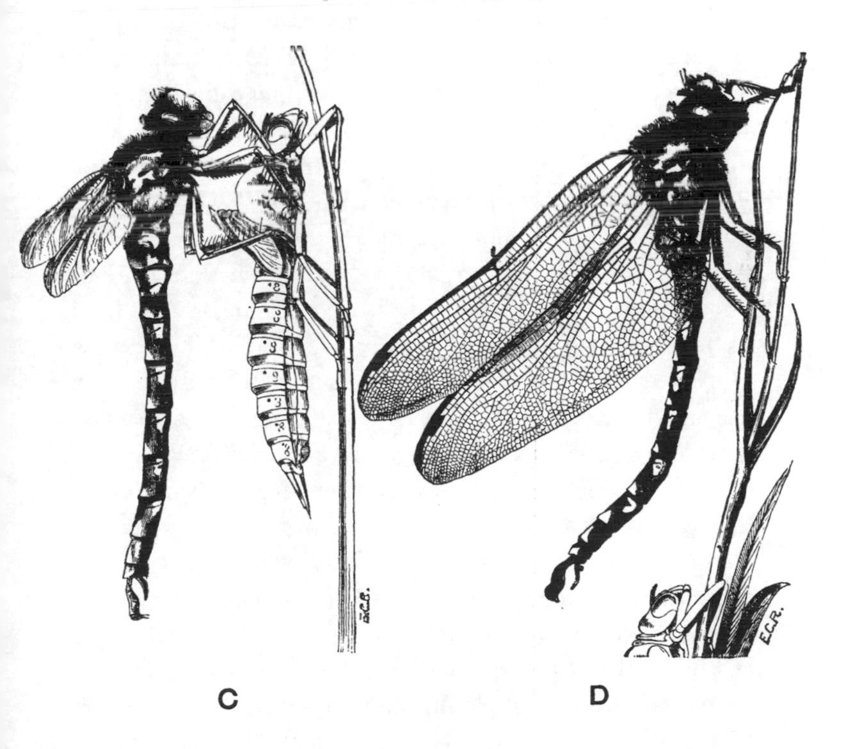

Fig. 19 (*continued*).

C. The whole body extricated. D. The perfect insect, the wings having acquired their full dimensions, resting to dry itself, preparatory to the wings being horizontally extended.

transparent. It discharges several drops of clear fluid from the anus. The fluid contains in solution nitrogenous compounds akin to urea. The colours of the body subsequently appear somewhat suddenly.

The wings however are as yet but little larger than those of the larva; they stand out from the back with the dorsal faces of right and left sides close together and the apices hanging downward. At this stage they are very soft and delicate and bleed freely if pricked or torn. The ensuing conduct of the freshly emerged imago appears to be designed for allowing the wings ample room for development, without risk of injury, and for exposure to a free current of air to aid in the drying and hardening. Often a few steps upward are taken so that a direct hold is obtained upon the support and the empty larval skin left behind. The abdomen is now bent ventrally, *i.e.* the imago "hollows its back," and the wings begin to expand. The process sets in at the base of the wing, and in about half-an-hour has extended up to the tip and the wings have reached their full size, but are still opaque and of a dingy yellow colour. Movements of the abdomen take place during this process and suggest the possibility of the expansion being caused by air being forced, in consequence of the contractions of other parts of the body, in between the two layers of which each wing is composed. The "hollowing of the back" affords space (in the *Libellulidæ*) for the development of the bellying and more fragile hinder part of the wing which would certainly be obstructed were not the abdomen moved out of the way by this bending. Some time, one or two hours, must yet elapse before the wings are dry and firm enough to support flight.

Emergence very often takes place in the early morning or may even begin before sunrise. It is clearly advantageous that the helplessness of the soft-bodied imago should be shrouded by the morning twilight from the eyes of insectivorous birds and other animals. Even when the wings are strong enough for flight, and the Dragonfly sails off through the air, the final appearance has not been reached, for there is a glossy sheen upon the surface of the wings themselves and the body lacks the full splendour of its colouring. This "immature" condition lasts one or two or several days, during which the insect often seems to avoid the ponds and streams. It is not known by what process or cause the "mature" appearance is assumed. There is evidence to show that, so far as the wings are concerned, it in some way depends upon the chief nervures being intact, for there is a record of a Dragonfly three of whose wings were "mature" while the fourth, which had suffered an injury to one of the principal nervures, and was therefore incapable of being used, remained "immature."

The life of the imago seldom lasts more than about three months. In England all die before the winter.

Enemies. As larvæ, especially when small and after each moult, Dragonflies probably fall frequent victims to fish; in aquaria they certainly devour one another and are attacked by predaceous aquatic insects, *e.g.* the larvæ of the large Water Beetle, *Dyticus marginalis.* The *imagines* are likewise at times guilty of cannibalism, the larger species catching their weaker brethren. A few birds have been observed to prey upon them, notably kestrels, hobbies and swallows, while Gilbert White and

Bechstein have noted cuckoos catching them both upon the wing and when at rest upon the reeds. I myself have seen the remains of Agrionids in the webs of Spiders (*Agelena labyrinthica*) spun in the gorse bushes beside a pond, and have often observed that the same weak-winged Dragonflies are caught and digested by the leaves of the Sundew (*Drosera rotundifolia*).

A small red mite, an Acarid, is occasionally (frequently in the genus *Sympetrum*) found as an external parasite upon the thorax.

CLASSIFICATION OF BRITISH DRAGONFLIES[1].

Dragonflies may be divided into two chief groups by the form and resting position of the wings.

(51) 1. Group I. ANISOPTERIDES. Front wings differ from the hind in shape, the latter being usually broader at the base: a small "accessory membrane" is present at the base of each wing on the posterior side close to the junction with the thorax. When at rest the wings are held spread out parallel with the dorsal and ventral surfaces of the body. The eyes meet on top of head (except *Gomphus vulgatissimus*). The 10th abdominal segment of the male has only one ventral appendage.

This group includes the two Families (A) *Libellulidæ* and (B) *Æschnidæ*.

They are distinguished from one another by the direction of a certain "cell," known as "the triangle," in the front wings.

[1] Lucas, *British Dragonflies*, London, 1900.

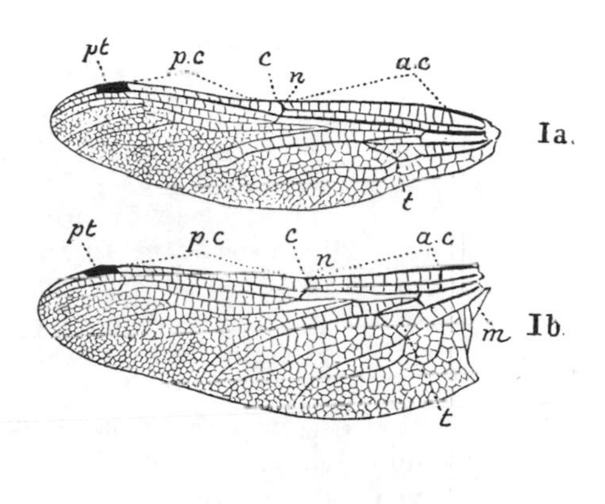

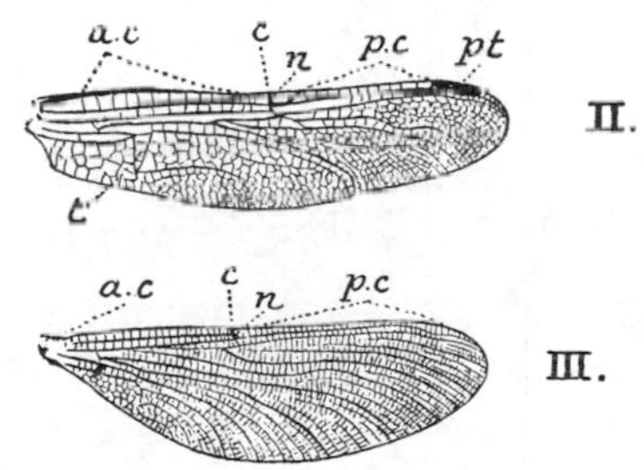

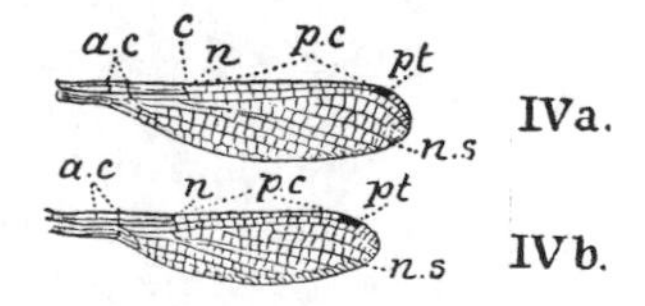

Fig. 20. Wings of Dragonflies.

Anisopterids. I a. Front wing, I b. Hind-wing, of an *Æschna*. II. Front wing of a *Libellula*.

Zygopterids. III. Wing of *Calopteryx* ♂. IV a. Front wing, IV b. Hind-wing, of an *Agrion*.

a.c. Antecubital nervures (transverse bars). *c.* Cubital point. *m.* Accessory membrane. *n.* Node (thick transverse vein). *n.s.* Nodal sector. *p.c.* Post-cubital nervures (transverse bars). *pt.* Pterostigma. *t.* Triangle.

Only those veins and cells referred to in the text are named.

(29) 2. Family A. LIBELLULIDÆ. The triangle of the fore-wing lies transversely to the length of the wing itself. Sub-families included are *Libellulinæ* and *Corduliinæ*.

(22) 3. Sub-family LIBELLULINÆ. 2nd abdominal segment of males simple, inner margin of hind-wing in this sex straight or convex, not concave. Appendages of 10th abdominal segment of male simple, of female cylindrical or fusiform, not flattened. Abdomen not bronzy in colour.

genera *Leucorrhinia*, *Libellula*, *Sympetrum*, *Orthetrum*.

(10) 4. Hind-wings with triangular dark spot at base.

(6) 5. genus *Leucorrhinia*. Abdomen with mid-dorsal spots, parallel-sided. Sole British species *Leucorrhinia dubia*, Lind. ♂ black with crimson spots, ♀ black with yellow spots. Length ♂ 35·5—39 mm., ♀ 34—36 mm. Expanse of wings ♂ 53—58 mm., ♀ 52—55 mm.

(5) 6. genus *Libellula*. Abdomen not spotted dorsally, sides not parallel.

3 species:

(8, 9) 7. *L. quadrimaculata*, Linn. Wings with a black spot at the node. Length ♂ 39—47 mm., ♀ 41—47 mm. Expanse of wings ♂ 64—74 mm., ♀ 70—80 mm.

(9) 8. *L. depressa*, Linn. Fore-wings with oblong dark patch at base, hind-wings with large triangular dark patch at base. Abdomen of ♂ steely blue, ♀ yellow-brown. Length ♂ 47 mm., ♀ 42—44 mm. Expanse of wings ♂ 75 mm., ♀ 75—78 mm.

(8) 9. *L. fulva*, Müll. Fore-wings with 2 dark lines at base, hind-wings with 1 dark line and small dark triangular patch at base. Abdomen of ♂ steely blue with black streaks, of ♀ tawny with blue-black central markings on hinder segments. Length ♂ 44 mm., ♀ 45 mm. Expanse of wings ♂ 74 mm., ♀ 75 mm.

(4) 10. Wings not spotted.

(19) 11. genus *Sympetrum*. Antecubital nervures about 7. Male red or black.

5 species:

(16) 12. Legs yellow.

13. *S. striolatum*, Charp. Dorsal surface of abdomen of ♂ red with yellow spots near hind margin of segments. Length of ♂ 39—44 mm., ♀ 39—42·5 mm. Expanse of wings ♂ 56—62·5 mm., ♀ 58—63 mm.

14. *S. fonscolombii*, Selys. ♂ abdomen crimson: nervures of wings of both sexes crimson. [A very rare visitor.]

15. *S. flaveolum*, Linn. ♂ abdomen dull red, base of wing in both sexes suffused with rosy yellow. [An occasional immigrant.] Length of ♂ 31—37 mm., ♀ 33 mm. Expanse of wings ♂ 48·5—60 mm., ♀ 56 mm.

(12) 16. Legs black.

(18) 17. *S. sanguineum*, Müll. Abdomen of ♂ crimson and dilated posteriorly, of ♀ yellowish olive with red mid-dorsal line. Pterostigma dark red in both sexes. Length of ♂ and ♀ 33—36 mm. Expanse of wings ♂ 50—57 mm., ♀ 53—58 mm.

(17) 18. *S. scoticum*, Don. Abdomen of ♂ almost black and dilated posteriorly, of ♀ yellowish brown, with black mid-dorsal line. Pterostigma black in both sexes. Length of ♂ 31—34 mm., ♀ 27·5—33 mm. Expanse of wings ♂ 48—51·5 mm., ♀ 42·5—52 mm.

(11) 19. Antecubital nervures about 12.

genus *Orthetrum*. Male abdomen blue in anterior part at least.

2 species:

(21) 20. *O. cærulescens*, Fabr. Pterostigma yellow. Abdomen of ♂ entirely blue, of ♀ yellowish brown. Wings of ♀ yellowish near base anteriorly. Length of ♂ 41—43·5 mm., ♀ 40—42 mm. Expanse of wings ♂ and ♀ 59—62·5 mm.

(20) 21. *O. cancellatum*, Linn. Pterostigma black. Abdomen of ♂ blue anteriorly, black posteriorly, of ♀ yellow with black lateral lines. Length of ♂ 48·5—52·5 mm., ♀ 49 mm. Expanse of wings ♂ 76—83 mm., ♀ 77 mm.

(3) 22. Sub-family Corduliinæ. 2nd abdominal segment of males provided with a pair of small projecting flaps—"auricles" and inner margin of hind-wing concave in this sex. Appendages of 10th abdominal segment of male

complex, of female flattened and leaf-like. Abdomen bronzy green in colour.

genera *Somatochlora*, *Cordulia*, *Oxygastra*.

(28) 23. Abdomen not spotted along mid-dorsal line.

(27) 24. genus *Somatochlora*. Pterostigma brown. Ventral appendage of 10th abdominal segment of ♂ triangular.

2 species:

(26) 25. *S. metallica*, Lind. Dorsal appendages of 10th abdominal segment of ♂ with one notch on ventral surface. Abdomen of ♀ slightly swollen at 2nd segment, then a little constricted, swelling out again very slightly posteriorly, colour metallic green, ♂ and ♀. Length of ♂ 52·5 mm., ♀ 56 mm. Expanse of wings ♂ 77 mm., ♀ 82 mm. [Very rare. Scotland.]

(25) 26. *S. arctica*, Zett. Dorsal appendages of 10th abdominal segment of ♂ with two or three notches on ventral surface. Abdomen of ♀ not swollen nor constricted, but gradually tapering to posterior end. Colour green with coppery-red tinge. Length of ♂ and ♀ 49 mm. Expanse of wings ♂ 67 mm., ♀ 68 mm. [Rare. Scotland and Ireland.]

(24) 27. genus *Cordulia*. Pterostigma black. Ventral appendage of 10th abdominal segment of ♂ cleft.

1 species: *C. æenea*, Linn. Colour metallic green; hinder part of thorax orange-brown. Length of ♂ 47·5—50 mm., ♀ 46—49 mm. Expanse of wings ♂ 66—70 mm., ♀ 68 mm.

(23) 28. Abdomen with yellow mid-dorsal longitudinal spots.

genus *Oxygastra*. Dorsal appendages of 10th abdominal segment with long spine on ventral surface at base; ventral appendage slightly notched at apex.

1 species: *O. curtisii*, Dale. Colour metallic green with yellow dorsal spots. Length of ♂ 53 mm., ♀ 50 mm. Expanse of wings ♂ 71 mm., ♀ 74 mm.

(2) 29. Family B. ÆSCHNIDÆ. Triangle of fore-wing placed lengthwise, parallel with long axis of wing. Sub-families included are *Gomphinæ*, *Cordulegasterinæ*, *Æschninæ*.

(32) 30. Eyes separate from one another.

31. Sub-family GOMPHINÆ. Females devoid of ovipositor. Ground colour black, with longitudinal yellow markings.

Sole British genus and species *Gomphus vulgatissimus*, Linn. Length 49 mm. Expanse of wings ♂ 62 mm., ♀ 68 mm., approximately.

(30) 32. Eyes meeting.

(34) 33. Sub-family CORDULEGASTERINÆ. Eyes meeting at a point only: females with ovipositor. Ground colour black with transverse yellow markings.

Sole British genus and species *Cordulegaster annulatus*, Latr. Length ♂ 75 mm., ♀ 85 mm. Expanse of wings ♂ 92 mm., ♀ 100—103 approximately.

(33) 34. Sub-family ÆSCHNINÆ. Eyes meeting for some distance: females with ovipositor. Genera *Anax*, *Brachytron*, *Æschna*.

(36) 35. Abdomen not spotted.

genus *Anax*. Abdomen with black mid-dorsal line of irregular width, ♂ abdomen blue, ♀ green: anal angle of hind-wings of ♂ rounded: no "auricles."

1 species: *A. imperator*, Leach. Length of ♂ 77 mm., ♀ 74—77 mm. Expanse of wings, ♂ 104 mm., ♀ 102—108 mm.

(35) 36. Abdomen spotted.

(38) 37. genus *Brachytron*. Abdomen with spots, some blue, some yellow. Thorax and anterior abdominal segments, dense, covered with pale hairs. Pterostigma very narrow. Anal angle of hind-wing of ♂ bluntly pointed. Ventral anal appendage of ♂ slightly notched at apex.

1 species: *B. pratense*, Müll. Length of ♂ 57·5 mm., ♀ 54·5 mm. Expanse of wings ♂ 71·5—74 mm., ♀ 74 mm.

(37) 38. genus *Æschna*. Abdomen variously spotted. Pterostigma of moderate width. Anal angle of hind-wings of ♂ distinctly pointed: ventral anal appendage of ♂ pointed.

6 species:

(48) 39. Abdomen nearly black, with blue or yellow spots. Wings clear, not coloured.

(42) 40. Eyes meeting for short distance only.

41. *Æ. cærulea*, Ström. Spots all blue in ♂, paler and inclined to yellow in ♀. Length of ♂ 66·5 mm., ♀ 58—64 mm. Expanse of wings ♂ 86—87 mm., ♀ 78—85 mm.

(40) 42. Eyes with long line of contact.

(46) 43. Dorsal blue spots on 9th and 10th abdominal segments divided in middle.

(45) 44. *Æ. mixta*, Latr. Costa of wing brown: suture between nasus and frons not black. Length of ♂ 61—66 mm., ♀ 62 mm. Expanse of wings ♂ 83—85 mm., ♀ 85 mm.

(44) 45. *Æ. juncea*, Linn. Costa of wing golden; suture between nasus and frons black. Length of ♂ 73·5—76 mm., ♀ 68—73·5 mm. Expanse of wings ♂ 94·5—99 mm., ♀ 93—97 mm.

(43) 46. Spots on 9th and 10th abdominal segments not divided.

47. *Æ. cyanea*, Müll. Length of ♂ 74 mm., ♀ 71—74·5 mm. Expanse of wings ♂ 102—104 mm., ♀ 106—110 mm.

(39) 48. Abdomen brown, hardly if at all spotted; wings brown or tawny.

(50) 49. *Æ. grandis*, Linn. Accessory membrane of hind-wing small, pale. 2nd abdominal segment with pair of narrow transverse yellow streaks dorsally. Eyes blue in ♂, brown with bluish sheen in ♀. Length of ♂ 70—76 mm., ♀ 69—76 mm. Expanse of wings ♂ 96—101 mm., ♀ 101—104 mm.

(49) 50. *Æ. isosceles*, Müll. Accessory membrane of hind-wing large, dark; 2nd abdominal segment with yellow triangle dorsally. Eyes green in ♂ and ♀. Length of ♂ 67 mm., ♀ 70 mm. Expanse of wings ♂ 91 mm., ♀ 94 mm. [Very rare. Eastern Counties. Early summer.]

(1) 51. Group II. ZYGOPTERIDES. Front and hind-wings alike in shape; no accessory membrane present; when at rest the wings are held more or less completely closed over the back parallel with the right and left sides of the body. The eyes do not meet on the top of head. The

10th abdominal segment of the male has two ventral appendages.

This group includes only the Family *Agrionidæ*.

Family AGRIONIDÆ. This being the only family representing the Zygopterids the characters of the group serve also for the family. Sub-families included are *Calopteryginæ*, *Agrioninæ*.

(55) 52. Sub-family CALOPTERYGINÆ. Antecubital nervures numerous: wings coloured, pterostigma of female white; male with no pterostigma. Sole British genus *Calopteryx*.

genus *Calopteryx*. (Characters of sub-family.)

2 species:

(54) 53. *C. virgo*, Linn. Wings of ♂ almost completely blue, of ♀ brown with bluish sheen. Length of ♂ 45—47·5 mm., of ♀ 43—46 mm. Expanse of wings ♂ 56·5—62 mm., ♀ 58·5—69 mm.

(53) 54. *C. splendens*, Harr. Wings of ♂ with blue patch across centre only, of ♀ greenish and pterostigma near tip of wing.

(52) 55. Sub-family AGRIONINÆ. Antecubital nervures only 2. Wings not coloured: pterostigma dark and in both sexes.

genera *Lestes*, *Erythromma*, *Pyrrhosoma*, *Ischnura*, *Platychnemis*, *Agrion*, *Enallagma*.

(65) 56. No blue spot behind eyes.

(62) 57. Abdomen not crimson.

(61) 58. genus *Lestes*. Abdomen of ♂ with segments 1, 2, 9 and 10 blue, remainder green. Eyes blue.

2 species:

(60) 59. *L. dryas*, Kirby. Rather robust: expanse of wings in both sexes considerably greater than length of body: pterostigma relatively broad. Dorsal anal appendages of ♂ with broad inwardly directed flange extending $\frac{2}{3}$ of length and bearing a narrow sharp tooth at about $\frac{1}{3}$ from base: distal $\frac{1}{3}$ of appendages broadly spatulate, turned sharply towards each other. Ventral anal appendages of ♂ barely half as long as dorsal and incurved. Black spots on dorsal surface of 1st abdominal segment

of ♀ subquadrilateral. Length of ♂ 36 mm., ♀ 35·5 mm. Expanse of wings ♂ 45 mm., ♀ 48 mm. [Very rare.]

(59) 60. *L. sponsa*, Hansem. Slender: wing expanse not much greater than length of body (generally), pterostigma relatively narrow. Dorsal anal appendages of ♂ with very broad inwardly turned flange extending rather less than ⅔ length and bearing a broad sharp tooth at about ⅓ from base and terminating abruptly, so as to give appearance of a 2nd tooth; distal remainder narrowly spatulate with distinct "handle," the pair turned gently towards each other. Ventral anal appendages of ♂ fully ¾ length of dorsal, parallel to each other. Black spots on dorsal surface of 1st abdominal segment of ♀ subtriangular. Length of ♂ 37—39 mm., ♀ 35·5—39 mm. Expanse of wings ♂ 40—44 mm., ♀ 43—47 mm.

(58) 61. genus *Erythromma*. Abdomen of ♂ with segments 1, 2, 9 and 10 blue, remainder steel black; of ♀ with 1st segment yellow with black spots, remainder black dorsally, yellow ventrally, anterior intersegmental sutures yellow, posterior blue. Eyes red.

1 species: *E. naias*, Hansem. Length of ♂ 34·5—35·5 mm., ♀ 35—35·5 mm. Expanse of wings ♂ 43—44·5 mm., ♀ 46—47 mm.

(57) 62. Abdomen crimson.
genus *Pyrrhosoma*.
2 species:

(64) 63. *P. nymphula*, Sulz. Thorax with a pair of longitudinal crimson stripes dorsally: posterior part of ♂ abdomen with dark bronze green markings. Length of ♂ 35—37 mm., ♀ 33—37·5 mm. Expanse of wings ♂ 43—46 mm., ♀ 46—50·5 mm.

(63) 64. *P. tenellum*, de Vill. Thorax without dorsal crimson stripes: ♂ abdomen entirely crimson; first 3 segments only of ♀ abdomen crimson, remainder bronze green. Length of ♂ 30—33 mm., ♀ 31—33·5 mm. Expanse of wings ♂ 33—36 mm., ♀ 36—39 mm.

(56) 65. Blue spot behind eyes.

(69) 66. Spot circular.

genus *Ischnura*. Abdomen black; part or all of segment 8 blue in ♂. Pterostigma more or less yellow; darker on front-wings than on hind of ♂. Nodal sector at level of 4th post-cubital nervure on fore-wings, of 3rd on hind-wings.

2 species:

(68) 67. *I. pumilio*, Charp. 9th segment and part of 8th blue in male: hind lobe of prothorax rounded and raised. Nervures of wings pale brown in both sexes. Length of ♂ 28—31·5 mm., ♀ 27·5—32 mm. Expanse of wings ♂ 30—34 mm., ♀ 33—37 mm. [Very rare.]

(67) 68. *I. elegans*, Lind. 9th segment black, whole of 8th blue in both sexes. Hind lobe of prothorax narrow, much raised. Nervures of wings dark. Length of ♂ and ♀ 30—33 mm. Expanse of wings ♂ 33—35 mm., ♀ 35—37·5 mm.

(66) 69. Blue spot behind eyes not circular.

(71) 70. Spot oblong. Mid and hind tibiæ dilated.

genus *Platychnemis*. Legs pale blue with fringes of black hair. Colour of body blue with black markings.

1 species: *Pl. pennipes*, Pall. Length of ♂ 36—37 mm., ♀ 36—38·5 mm. Expanse of wings ♂ 43—46 mm., ♀ 45—48 mm.

(70) 71. Blue spot behind eyes pear-shaped: tibiæ not dilated.

(77) 72. genus *Agrion*. Abdomen blue and black. Pterostigma bluish with black centre, alike on both wings. Nodal sector level of 5th post-cubital on fore-wings and of 4th on hind-wings. ♀ with no apical spine on ventral surface of 8th segment.

3 species:

(76) 73. U-shaped mark on second abdominal segment of ♂.

(75) 74. *A. pulchellum*, Lind. ♂ with dorsal U-shaped black mark on 2nd abdominal segment united to black ring behind it: ♀ blue and black. Length of ♂ 32—36·5 mm., ♀ 34—36 mm. Expanse of wings ♂ 36—41 mm., ♀ 44—50 mm.

(74) 75. *A. puella*, Linn. U-shaped black mark on 2nd abdominal segment of ♂ free from black ring behind it: ♀ black-

bluish-green with narrow yellowish intersegmental rings. Length of ♂ 33—36 mm., ♀ 34—36 mm. Expanse of wings ♂ 36—40 mm., ♀ 44—48 mm.

(73) 76. *A. mercuriale*, Charp. ♂ with black mark shaped like symbol for the planet Mercury upon 2nd abdominal segment (a crescent, with horns turned forward, standing on a "mushroom" whose stalk joins black ring behind). Pear-shaped blue spot behind eyes nearly circular in both sexes. ♀ abdomen black-bronze: ring between 7th and 8th, and 8th and 9th segments clear blue. Length of ♂ 27—31 mm., ♀ 29—31 mm. Expanse of wings ♂ 31—35·5 mm., ♀ 35—39 mm. [Rare. New Forest.]

(72) 77. genus *Enallagma*. Abdomen and nodal sector as in *Agrion*. Pterostigma entirely black, or with pale centre, alike on both wings. ♀ with apical ventral spine on 8th abdominal segment.

1 species: *E. cyathigerum*, Charp. With characters of the genus.

CHAPTER V.

WASPS.

WITHIN the limits of the Order Hymenoptera are to be found many of the most interesting and highly gifted insects of our British fauna. The "Saw-flies," "Ichneumon-flies" and "Gall-flies" constitute one section of the Order and are characterised by possessing no sting, but in its place a boring organ for piercing the surface of the plant or animal within which the eggs are deposited. It is however in the Aculeate (Sting-bearing) section that the greatest complexity of social organisation is attained. In the lower members of this section, such as the *Fossores* and *Sphegidæ* (Sand Wasps), we already find the maternal instinct strongly developed. The females dig holes in the ground, or burrow into wooden posts, bramble stems, etc., and there construct one or more cells. In each cell is placed a single egg and a store of animal food for the nourishment of the future grub. Spiders, dipterous flies, caterpillars, *Aphides* and many others are thrust alive, but paralysed by the poison-sting of their captor, into these nursery-larders.

The domestic economy of the lower bees (*Andrenidæ*, etc.) is not greatly in advance of the above; their chief claim to superiority rests in the fact that it is vegetable food (bee-bread) gathered as pollen from flowers that they lay up for their offspring. In the ants, the honey bee, bumble bees and wasps (*s. s.*) however we meet with the elaborate social conditions of vast communities toiling with one object—the welfare of the race. Many individuals in each society devote their entire energies to the work of building, foraging or nurturing the young, but—so complete is the division of labour—take no direct part in propagating the species. These sterile individuals—"Workers"—may even exhibit differences among themselves in accordance with their respective duties. In any case these Social Hymenoptera comprise males (Drones), fertile females (Queens) and sterile females (Workers).

There are seven (six if *V. rufa* and *V. austriaca* are but varieties) species of Social Wasps (genus *Vespa*) found in this country. Inasmuch as they are armed with a poisonous sting it has proved advantageous to advertise the dangerous quality to insectivorous foes. Accordingly wasps are conspicuously marked with yellow and black patterns. The seven species closely resemble each other and thus confer mutual benefit upon each other by sharing the losses sustained through trial-tasting on the part of inexperienced animals, *e.g.* young birds. It has been established beyond dispute that young insectivorous creatures do not instinctively know, but have to learn by painful experience (one lesson is usually enough), that conspicuously coloured insects are as a rule dangerous or unpalatable. There are however exceptions, for a considerable number of defenceless insects mimic the black and yellow uniform of the wasps and, sailing under false colours, escape attack. Of our English insects which mimic wasps we may mention in this category the dipterous flies —*Chrysostomum sylvarum*, *Volucella inanis*, *V. inflata* (in a less degree) and *Myriatropa florea*; the beetles *Strangalia armata*, *Clytus abietis*, *Rhagium bifasciatum*, *Pachyta octomaculata*, *Necrophora ruspator*, *N. vespilio*, *Cicindela sylvatica*, *Callistus lunatus* and others; and the Clearwing Moths of the genus *Sesia*.

It may be well to point out how a wasp may be known and with certainty distinguished from its mimics without having recourse to the sting. Wasps, then, have two pairs of transparent membranous wings; when at rest they fold the anterior wings longitudinally so as to diminish their breadth by one-half. Wasp-like flies have only one pair of wings and these they do not fold; the beetles have only the hinder wings membranous, the front pair being horny and opaque; Clearwing Moths have two pairs of more or less transparent membranous wings but they do not fold them when in repose. The social wasps (*Vespidæ*) may be known from their near relatives the solitary wasps (*Eumenidæ*) by the elongated longitudinally grooved mandibles and by the bifid or toothed claws of the latter: a Vespid also possesses two spurs at the apex of the middle tibiæ whereas an Eumenid has but one.

In powers of locomotion wasps are, on the whole, superior both to the dragonfly and cockroach. The cockroach is a swift runner but an infrequent flier, the dragonfly very powerful on the wing but uses its legs very seldom for walking: the wasp flies strongly and can travel at a fair pace by walking or running. The mechanism of the wings is such that the anterior and posterior wings of the same side move as one. At about half-way along the anterior border of the hind-wing is a row of hooks (4 Fig. 21) (thirty-two in number in the females, but fewer, generally about twenty, in the males),

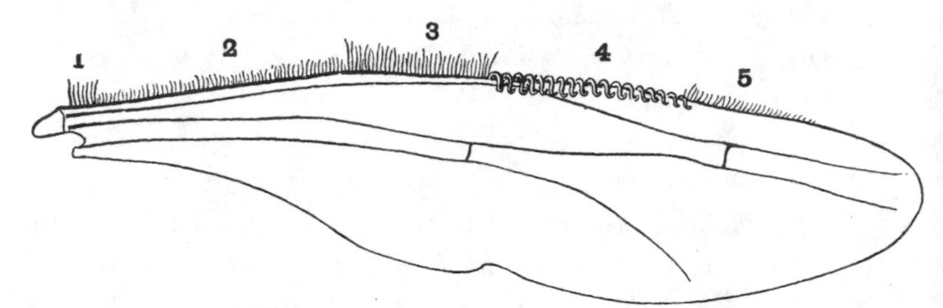

Fig. 21. Hind-wing of a wasp. Magnified.

1. Tuft of strong hairs. 2. Short hairs. 3. Long hairs, all of which project forwards under the front wing. 4. Hooks for fastening anterior edge of hind-wing to posterior edge of front wing —only a few hooks are represented. 5. Short hairs.

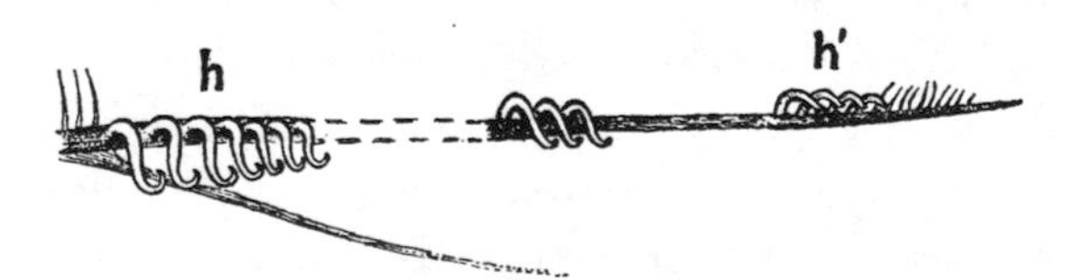

Fig. 22. Part 4 of Fig. 21 to show the difference between the shapes of the proximal (*h*) and of the distal (*h'*) hooks.

with slightly upturned points. On the posterior edge of the front wing is a crest forming a ledge, beneath which the ends of the hooks catch when, and as, the wings are extended for flight. The hooks nearest to the attachment of the wing are the largest and their tips are curled over; but they gradually get smaller as they pass towards the extremity of the wing and the tips become straight: at the same time they stand out further from the margin of the wing (*h'* Fig. 22). In addition to the hooks there is at the base of the hind-wing a strong tuft of hairs (1 Fig. 21), and along the margin at first short hairs (2) and then longer hairs (3) projecting forward (the hooks are merely modifications of these hairs). These all are pressed up against the under surface of the fore-wing and assist in keeping the two wings together.

This method of interlocking the fore and hind-wings is usual in Hymenoptera, but other contrivances for the same purpose are found among other insects. For example, most Lepidoptera have a stiff bristle or group of stiff hairs (the *frenulum*) arising from near the base of the anterior edge of the hind-wing and engaging in a chitinous catch or group of stiff scales (the *retinaculum*) on the lower surface of the front wing. In others, *e.g. Endromis versicolor*, the Kentish Glory, members of the genera, *Eriogaster*, *Clisiocampa*, *Odonestis*, the eggars, lackeys, fox-moths, drinkers, etc., there is no definite interlocking arrangement, but the basal area of the anterior part of the hind-wing is expanded so as to underlap the front wing for some distance and rest more firmly against it, thus avoiding dislocation. The *Hepialidæ* (Swifts) and *Micropterygidæ* present yet a third system among the Lepidoptera. In these a membranous or a spine-like process (*jugum*) springs from the hinder edge of the fore-wing near its base and passes under the hind-wing, which is thus held between the jugum and the overlapping part of the fore-wing. Instances might be multiplied from other orders of insects.

It is to be noticed that the interlocking of the wasps' wings is effected in such a way as to not only hook them together but also to keep them parallel and in contact during motion. The hooks do not fit into "eyes" but slide upon a bar, and are not disengaged and put out of gear until the wings are brought into the position of rest. Upon the surface of the wings are numbers of short stiff hairs pointing obliquely backward and to the tip of the wing: these catch and get a grip upon the air during the downstroke, but upon the upstroke allow it to slip off.

The musculature of the thorax[1] (which in this Order includes the first abdominal segment also) is very complicated, inasmuch as this region of the body contains not only all the muscles for the movement of the wings and legs, but also others for the movement of the skeletal parts of the thorax itself. It is to the necessity of affording relatively rigid attachments for all these muscles that the firm, hard exoskeleton of the thorax is due. Passing over the smaller muscles which are concerned with extending the wings for flight or bringing them back into the position of repose and making minor adjustments of parts involved we will attend only to those which bring about the vibration of the wings. Janet[2] has shown that the hind-wings are destitute of muscles and remain motionless except their

[1] Chabrier, "Vol des Insectes," *Mem. Mus.* 6, 7, 8, 1820. Bennett, *Zool. Journ.* II.

[2] *C. R. Acad. Sci.* Paris, CXXVIII. 1899.

hooks have become engaged in the ledge of the fore-wings, in which case they partake in the movements of the latter. It is therefore only necessary to consider the motor mechanism of the fore-wings. There are no muscles attached directly to the wing itself, but its vibratory movements are brought about by alterations in the shape of the framework of the meso-thorax. These changes are effected by two sets of powerful muscles; a longitudinal set from the anterior to the posterior end of the meso-thorax occupying the dorsal two-thirds of the median region, and a paired transverse set, to right and left of the first, running nearly vertically from the dorsal to the ventral surface. The articulation of the extended wing may, for the purpose of understanding the principle of the mechanism, be regarded as a plate at right angles to the wing and situated about half-way along the side of the meso-thorax at the ventral end of the hinge between the dorsal parts of the anterior and posterior portions of the meso-thorax (*scutum* and *scutellum*). This "wing-plate" must be considered as firmly attached to both scutum and scutellum. When the longitudinal muscles contract they draw the anterior part of the scutum and the posterior part of the scutellum towards each other. This shortening of the meso-thorax results, in consequence of the above-mentioned hinge and of the thrust against certain parts of the internal skeleton, in the sides of the hinge getting pinched together and arching in a dorsal direction; *i.e.* the meso-thorax becomes deeper dorso-ventrally. The hinge in its dorsal movement draws with it the "wing-plate," throwing the latter through about a right angle from an almost horizontal into a vertical position. This is possible in virtue of the "wing-plate" lying in a recess or fold. It must be borne in mind that when the "wing-plate" is horizontal the wing itself, being set at right angles to the plate, is vertical and *vice versâ*. Hence when the longitudinal muscles contract the wing is depressed—the down-stroke is made. The up-stroke is brought about by the contraction of the transverse dorso-ventral muscles which pull the scutum and scutellum down again, separating the sides of the hinge and at the same time lowering the "wing-plate" and restoring it to the horizontal, but the wing to the vertical position.

The action may be roughly imitated as follows:—crease a strip of fairly stiff paper as in Fig. I. and push a pin through *BC*: then the pin *P* represents the wing, *BC* = "wing-plate," *D* = dorsal and *A* = ventral surface of meso-thorax. *BC* is nearly horizontal, *P* nearly vertical. Now hold *BA* firmly and pull *D* sharply upwards, imitating the effect produced by the longitudinal muscles; the crease opens out, *BC* becomes vertical, the point of *P* (wing) travels over an arc downwards into the position shown in the second figure; this completes the down-stroke. The up-stroke is made by pressing *D* down, when the crease becomes folded once more and the point of *P* travels up, *BC* resuming the nearly horizontal attitude.

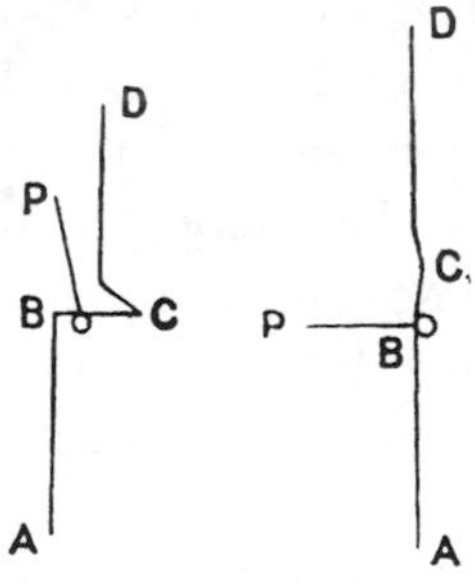

Fig. I. Fig. II.

If a wasp be held firmly but allowed to vibrate the wings freely, the movements of the dorsal surface of the meso-thorax can be readily detected by means of a long and light lever rested upon it. Moreover a very slight dorso-ventral pressure upon the meso-thorax completely arrests all movement of the wings.

It must not be imagined that the actual movement of the wing is so simple as this description indicates. Marey[1] and Pettigrew have independently pointed out that the extremity of the wing describes an 8-shaped trajectory during one complete oscillation. The former of these investigators has made it clear that, though it is true that the muscles maintain only the up-and-down movement, the resistance of the air operating upon the more flexible hind portion of the wing induces numerous changes in the inclination of the surface of the wing and thus determines the form of the trajectory.

The number of vibrations per second is in a wasp about 190. This fact has been determined by causing the wing of a captive insect to record its strokes upon a rapidly revolving cylinder whose surface is blackened with smoke. A tuning-fork, whose number of vibrations per second is known, is made simultaneously to record a tracing on the same cylinder. Thus a direct comparison is possible between the two series of markings.

The rapid movement of the wings is in part responsible for the well-known buzzing of wasps and other insects. The slight change in the note heard as an insect is flying is not due to any alteration in the frequency of vibration,

[1] *Movement.* Trans. by E. Pritchard. Heinemann, 1895.

but depends upon whether the flight is directed towards or away from the observer. The note becomes higher as the insect approaches and lower as it recedes. A precisely similar phenomenon, due to the same cause, may be noticed in the whistle of an express train approaching, rushing through and receding from a railway station. If an insect be held captive when its wings vibrate the note produced is constant. According to Pérez and Bellesme[1] there are two distinct sounds in the buzzing. One, a deep noise, is due to the vibration of the wings and is produced whenever a certain rapidity is attained; the other is a shriller sound and is said to be produced by the vibrations of the walls of the thorax to which muscles are attached. Both of these observers agree that the spiracles are not, as has been maintained by some, concerned in producing the sound.

Shipley and Wilson[2] have described an apparatus on the wing of the mosquito to which the high-pitched note of the buzz of this insect may be due. The organ lies at the extreme base of the wing very close to the articulation with the thorax. It consists of a slightly movable bar *A* (Fig. 23), which bears on its hinder, free edge a series of well-marked teeth from thirteen to fifteen in number. Posteriorly to this bar there is situated a chitinous blade *B*, with from thirteen to fifteen sharply defined and slightly oblique elevations. The teeth of the bar *A* rasp against the ridges of the blade *B* and a shrill note is

[1] *C. R. Acad. Sci.* Paris, LXXXVII. 1878.

[2] On a possible stridulating organ in the mosquito (*Anopheles maculipennis*, Meig.), *Trans. Roy. Soc. Edinb.* XL. pt. ii. No. 13, 1902.

thereby produced. It is noteworthy that the sound can be produced by the insect when the wing has been reduced by amputation to a mere stump. As more and

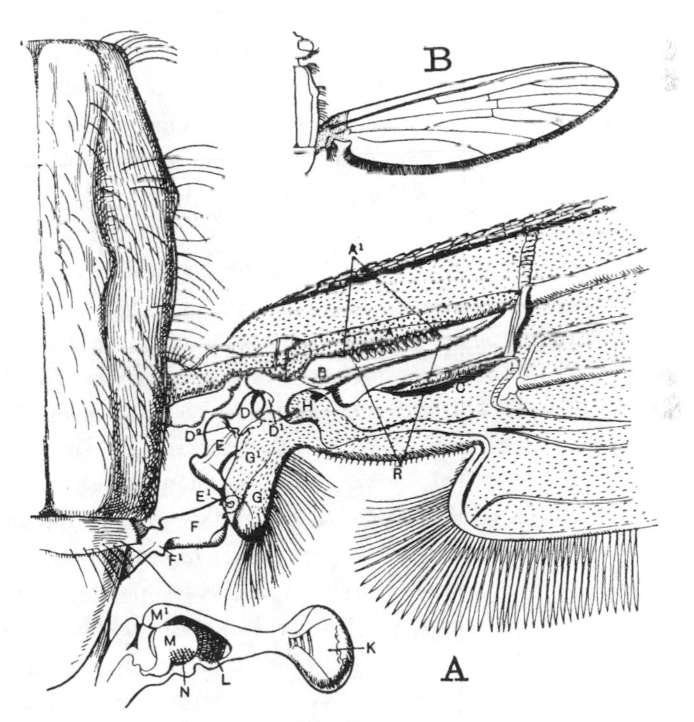

Fig. 23.

A. Right half of thorax of *Anopheles maculipennis*, Meig., with base of right wing and right halter. Magnified about 45.

B. The same magnified about 7 to show the area which bears the stridulator. This area is shaded.

A. Toothed bar. A^1. The teeth which rasp on the ridges borne by *B*. *B*. Blade bearing the ridges *R*. *C*. Trough which limits the movements of *B*. *K*. Distal end of halter. The remaining letters refer to details not mentioned in the text.

[From A. E. Shipley and E. Wilson.]

more of the wing is cut off the volume of the sound decreases but the note rises in pitch.

There is no doubt that the buzzing produced by many insects has a protective value in warning insectivorous foes of dangerous attributes. Lizards and frogs which have once had experience of the sting of a bee exhibit serious alarm at the loud buzz of the harmless mimic *Eristalis tenax*, the drone fly. I have noticed that the sand-wasp, *Ammophila hirsuta*, emits a loud and angry buzz while underground engaged in excavating her burrow, and have observed would-be inquiline Diptera and Hymenoptera start away from the hole and retreat hastily on hearing the sound emerge from the ground.

The functions of the legs, digestive, circulatory, respiratory, nervous and excretory systems of the wasp do not greatly differ from those of the cockroach described in an earlier chapter. We will therefore pass on to consider that most distinctive appendage of the higher Hymenoptera, the sting. Strictly this organ belongs to the female reproductive system and corresponds to an ovipositor. It has however been withdrawn into the body and lost all direct connexion with oviposition, the eggs no longer passing through it but escaping at its base. Its existing function is that of a weapon of defence and offence. The males are of course destitute of a sting. The sting[1] itself consists of three long pieces, a larger dorsal "director" (*D*) (usually mistaken for the sting itself)

[1] Cf. Kraepelin, *Zeit. wiss. Zool.* XXIII. 1873. Carlet, *Bull. Soc. Ent. Fr.* (*b*) IV. 1884 and *C. R. Acad. Sci.* Paris, XCIX. 1884. Dewitz (Bee), *Zeit. wiss. Zool.* XXV. 1875. (Ant) *ibid.* XXVII. 1877.

and two three-sided "needles" bearing six barb-hooks at their apices. The "director" tapers posteriorly and is rounded dorsally so as to form rather more than a half-cylinder: on its ventral face is a deep groove in which the needles slide to and fro: and the edges (*E*) of the groove are turned in so as to receive the outer ridges of the needles right and left. The whole may be compared to an inverted gutter-pipe with a flange turned in partly closing the concavity and lodging two rods within it (*vide* Fig. 24). The needles can thus slide freely up

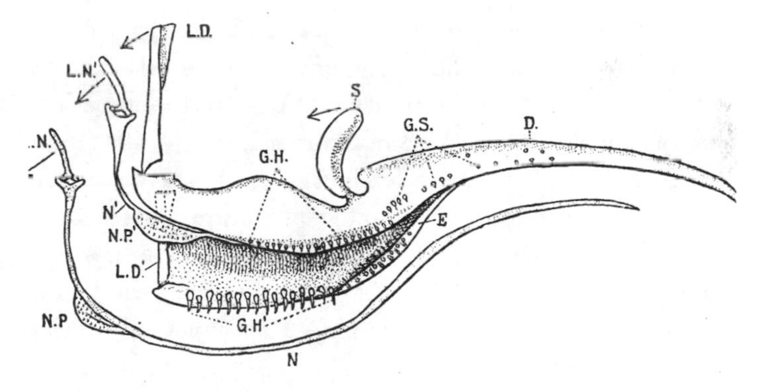

Fig. 24. Ventro-lateral view of sting of *Vespa germanica* ♀: the needle of the far (right) side has been withdrawn and is shown below the other structures.

D. Director. *E.* Ventral inturned edges of director. *G.H*, *G.H'*. Guiding hairs of left and right sides. *G.S.* Guiding studs on inner face of *D.* *L.D*, *L.D'*. Levers of left and right sides for extrusion of *D.* *L.N'*, *L.N.* Levers for extrusion of left and right needles, *N'*, *N*, beyond tip of *D.* *N.* Right needle withdrawn from *D* but lying parallel to its original position. *N'.* Anterior part of left needle; the remainder lies concealed within *D.* *N.P*, *N.P'*. Piston (?) enlargements on right and left needles. *S.* Stout chitinous piece on dorsal surface of *D.* The arrows show the direction in which the muscles pull the levers when the sting is employed. The poison sac and duct are not shown.

and down within the director but cannot be separated from it nor drop out from the groove.

At its anterior end the "director" is enlarged and its groove widened: at this point the needles become free from the inturned flange but are still kept in contact and parallel with the margin of the "director" by a double series of peculiar hairs (*GH*), about 50 in number, situated on its inner face near the ventral edge. The outer edge of the needle runs between the upper and lower series of hairs and is thus retained in position. Within the tapering part of the director the hairs are replaced by studs (*GS*). These cluster in a group at first, but in the narrower parts of the director are isolated and rather irregularly disposed. They serve to guide the needles in their passage through the director. There is no continuous "bead" such as Carlet describes as performing this function in the sting of the honey-bee.

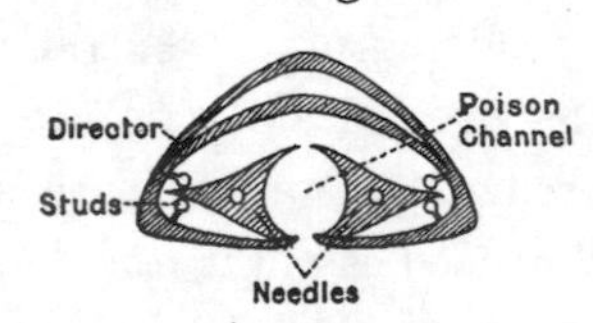

Fig. 25. Diagrammatic cross-section of a wasp's sting in the region of the letter *D*, in Fig. 24.

Anteriorly the needles diverge from one another and bend rather sharply in a dorsal direction. Between them is interposed a bulbous muscular mass (not shown in the figure) attached to the anterior end of the director. Finally the needles are articulated to strong levers *LN*, *LN′* which give attachment to powerful muscles. When the handle of the lever is pulled down by these muscles, in the direction shown by the arrows in the figure, the arm which articulates with the needle is thrust backward and slightly upward and the needle itself is darted along the groove of the director and projects beyond it, escaping at a spot just short of the extremity.

In the act of stinging the first incision is made by the point of the director, which is itself darted out by the action of levers *LD*, *LD′* and muscles: on its dorsal surface is a strong piece, *S*, which at first holds the director down, but eventually, when full extrusion has been reached, causes the director and needles to turn abruptly upwards towards the dorsal surface; then by a rapid alternate movement of the levers the needles are driven deeper into the wound. The two concave inner faces of the needles are pressed firmly against each other, partly by the tapering sides of the director and partly by the course imparted by the muscles, and so form a closed tube between them. Down this tube the poison is driven by the contractions of the muscular wall of the bag in which the poison is stored. An enlargement *NP*, *NP′* on each needle near the anterior end perhaps acts as a piston on reaching the narrower part of the director: the two pistons virtually sweep the poison in front of them. Their surfaces are covered with numerous fine scales like those of fish. The two together may serve to close the triangular cleft between the edges (*E*) of the director when the needles have been thrust out. In this position they would prevent the liquid poison from escaping through the triangular cleft. It is possible that they perform both duties.

The duct from the poison bag discharges close against the anterior enlargement of the director between the diverging anterior portions of the needles which are thus conveniently placed for receiving the fluid between them. The poison escapes from between the needles at their apices and also through five minute canals *pc* (Fig. 26), that pass obliquely from the poison groove through the bases of the first five barbs *b* (Fig. 26). Neither the director nor the needles are solid structures; tracheal tubes are plainly visible within their substance and distinct cavities are visible in transverse sections.

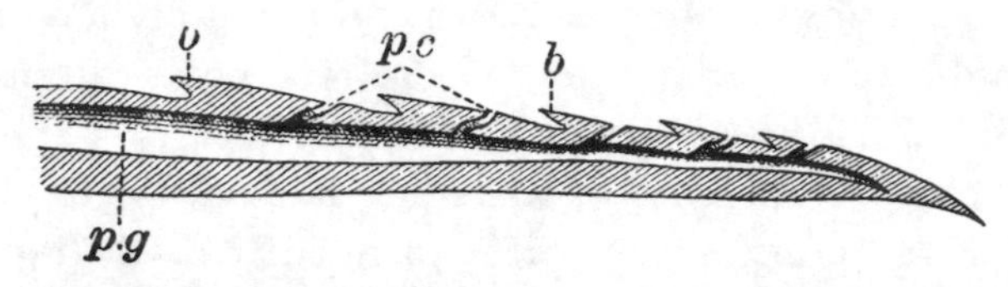

Fig. 26. Optical section of tip of a needle from the sting of *Vespa germanica* ♀. Highly magnified.

b. Barbs. *p.c.* Poison canals perforating bases of barbs. *p.g.* Poison channel running along inner face of needle, cf. Fig. 25. Actual length of the part figured 0·37 mm., breadth at widest part 0·05 mm.

The poison is said by Bordas[1] to be formed by the mixture of the secretions of two glands, of which one is acid and the other alkaline. It certainly gives a strong acid reaction with blue litmus paper, but we have not been able to detect the alkaline constituent.

It may perhaps be not out of place to observe that the painful effect of a wasp's sting is entirely prevented by the prompt application of ammonia to the wounded spot. The popular idea that a wasp is always anxious to use its sting is entirely wrong: so long as the insect is not interfered with and hasty movement is avoided there is not the least likelihood of the weapon being employed. It is

[1] *C. R. Acad. Sci.* Paris, cxviii. 1894.

perfectly possible to stand unhurt, and indeed unnoticed, within a few feet of a wasp's nest in full activity so long as all movements are made without flurry or haste.

Life History[1]. The life history of the individual wasp takes its appropriate place in that of the colony. Each "nest" is founded and populated by but one "queen." In the early spring the hibernated "queens" are roused to activity by the warmth of the atmosphere and issue forth from their winter quarters. These have not been found in the old nests of the last season, but each "queen" after mating in the previous autumn seeks for herself a secluded and sheltered spot in such places as the thatch of a roof or rick, under sacking or loose rubbish, or even in the folds of curtains, etc., in our houses. There she seizes some shreds between her jaws, draws her legs up close to the body, making no use of her claws for purposes of support, but relying solely upon her jaws. The antennæ are turned down and held under the first pair of legs; the wings are brought down on the ventral side of the body between the third pair of legs and the abdomen. In this position she sleeps for some months. It is noteworthy that the legs, wings and antennæ are brought once again into the position which they occupied during a previous period of inactivity, namely, during the pupa stage, the only difference being that the wings are now of greater size.

[1] Cf. Paul Marchal, *Arch. de Zool. Expér. et Gén.* (3) IV. 1896; C. Janet, *Mém. Soc. Zool. Fr.* VIII. 1895; *Mém. Soc. Acad. de l'Oise* XVI. 1895; *C. R. Acad. Sci.* Paris, CXX. 1895; *Études sur les Fourmis, les Guêpes et les Abeilles*, Limoges, 1895.

When once hibernation has set in a very great degree of cold can be endured with immunity. I have known wasps exposed to a temperature of 10° F. (22° F. below the freezing point) without any ill effects. On the other hand unseasonable warmth awakes them and they are then liable to disaster. I have seen "queen" wasps on the wing in the open on December 26th and on February 7th. Hence a severe winter which keeps them dormant is favourable and likely to be followed by an abundance of wasps in the following summer, while the reverse is the case if the winter be mild with a few intermittent "snaps" of hard weather.

The first business of the duly awakened "queen" is to find a spot suitable for her nest. Our commonest English species build "underground," and for these a deserted mouse-hole, a crevice in a wall, the thatch of a barn and other similar places are eligible building sites. Prospecting "queens" may often be seen in April and May searching along hedge-banks and walls for a place to their liking.

When suited the "queen" takes notes of the surroundings so as to know her bearings and to be able to return home. She then proceeds to gather material for the construction of her nest. This she obtains by rasping off with her jaws the weathered surface of wooden posts, palings, etc.: rotten wood is not used. Nearly any oak fence in the open country bears upon it hundreds of marks as if it had been lightly scraped with the finger nail; these marks are made by the jaws of wasps. A fair sized pellet having been collected and moistened by saliva from her mouth, the "queen" carries it off to her chosen house and applies it as a thin layer of "wasp-paper" to the top of the cavity which is to hold the nest. A rootlet or rafter or a well-placed straw in the thatch usually serves for the foundation.

By repeating the above process at length a disc is made, and from the centre of this a narrow stalk is hung. At its lower end the stalk widens out and here are laid down the outlines of the first four cells. There is no one central cell but the rest of the comb is added round these four which are arranged thus—

The next four are added as shown, and so on with the succeeding cells up to about thirty. The cells are closed above but open below. An umbrella-shaped cover is hung from the foundation to protect the comb and its contents. As soon as each cell is outlined and provided with a low inverted parapet—the side walls—an egg is deposited in it, being fixed by a cement in the angle nearest to the centre of the comb.

In a few days, the exact length of time depending upon the temperature, a legless grub emerges from the egg. The grubs possess jaws and are fed chiefly on the juices of animal food such as caterpillars, *Aphides,* and flies, by their mother. As they grow they moult their skins periodically. The cells are made longer by the queen as the increasing size of the grubs demands. The grubs are placed head downwards in their cells so that their mouth is at the open end. They at first retain their position and avoid falling out by means of the cement by which the egg is glued to the cell wall, for the hinder segments remain in a portion of the egg-shell. Subsequently the curved shape of the body causes it to abut firmly on both dorsa. and ventral sides against the walls of the cell.

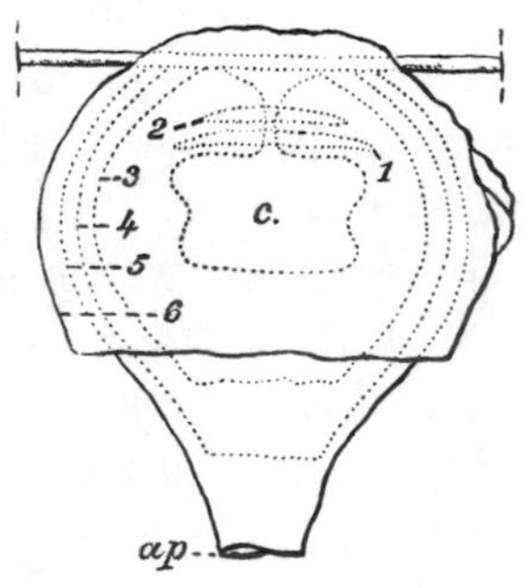

Fig. 27. Diagram of wasp's nest at an early stage. (After Janet.)
ap. Entrance to nest. *c*. Comb. 1, 2. 1st and 2nd coverings now largely removed to make room for growing comb. 3, 4, 5. Corresponding coverings still complete. 6. 6th covering as yet not finished.

When full grown the grub spins a cocoon within the cell, lining the entire cavity sparsely, but closing the bottom with a tough silken floor which projects as a dome beyond the lower end. In this operation the grub completely reverses its position, turns, in fact, head over heels in order to apply silk discharged at the anterior end of the body to all parts of the cell wall. Hitherto no fæces have been extruded: now, however, a black mass, composed of a sac containing the entire fæcal accumulations of larval life, is ejected. This mass remains attached to the top of the cell and becomes flattened by the pressure of the larva against it. A marked change now takes place in the form of the animal. A moult occurs shortly after the completion of the cocoon, and the pupa (chrysalis) stage is reached. Instead of a legless grub we now find a creature provided with distinct head, thorax and abdomen and their respective appendages, viz.: antennæ, jaws resembling those of a mature wasp, three pairs of legs and two pairs of wings. These appendages are folded against the body and the whole is of a semi-transparent white colour. It differs from the familiar pupa of a butterfly or moth only in the fact that the parts do not become firmly glued together and hard, but remain soft and easily separable from one another. In short it persists in the condition that is found for the first half-hour or so of pupal life in a Lepidopterous insect. No food is taken during this period, but elaborate internal changes and re-arrangements are effected at the expense of the stores of fat already accumulated within the body of the grub-larva. It is not until several days have passed, and the pupal period is drawing to a close, that any power of movement is manifested even though the pupa be extracted from its cocoon. Gradually the *imago* or perfect wasp becomes visible through the skin of the pupa which at length bursting open sets free from its shroud the head and eventually the entire body.

The young *imago* soon attacks the floor of the cocoon with her mandibles, bites a way through and crawls out upon the lower surface of the comb. The development from the egg to the *imago* occupies about a month or six weeks, according to the temperature. A spell of cold wet weather much delays progress and may prove fatal to the larvæ and the entire community. As yet the body is moist and its colours dull, the wings too are not fully expanded nor firm. For some time the newly emerged wasp rests upon the comb, hanging back downwards, while the wings attain their final form and consistency and the body its strongly contrasted colouring. Not infrequently rest is taken by entering a cell, her own or some other, head first. During this period there are ejected from the body the nitrogenous waste-products that have accumulated within the hinder part of the intestine during the period of quiescence. These excreta take the form of drops of milky fluid and small white granules largely composed of urates. A similar discharge of waste-products is generally very noticeable in freshly emerged Lepidoptera: in these the colour of the liquid is pink or red and has given rise at times to assertions of "showers of blood" among country folk.

The fore-wings of the young wasp are not, at first, folded longitudinally, but overlap the dorsal surface of the abdomen, and for a slight distance one another also. They do not become folded until they have been extended and become hooked on to the hind-wing: then, on returning to the position of repose, this attachment is maintained and the hind part of the fore-wing folds under the front part mechanically. Janet has shown that in the absence of a hind-wing the fore-wing of that side does not, and indeed cannot, become folded. He also suggests that by this folding the more fragile parts of the wings are preserved from the injury they might otherwise receive by friction between the abdomen and the fabric of the nest.

One of the first acts of the wasp in full power of her limbs is to clean herself with the brushes and combs already described (see p. 77). Then she walks about upon the comb, visiting cells which contain large larvæ and touches their heads with her jaws. Each larva so visited gives out of its mouth a drop of liquid which the young wasp eagerly swallows. For a day or two the young "worker," for such she is, remains within the nest and begins to relieve the "queen" of some of her labours. When the latter returns to the nest with a load of food, the worker takes it from her and distributes it to the grubs.

After a couple of days the young worker issues from the nest and sets about the task of collecting food for the grubs and fresh material for the enlargement of the nest. So soon as the "queen" has reared about a dozen helpers she gives up to them the entire care of the grubs and of building. Henceforth she remains within the nest and devotes herself solely to depositing eggs in the cells prepared for her.

Meanwhile the external appearance of the nest has been changed. The umbrella-shaped covering is prolonged by the "queen" and made balloon-like with a circular hole at the bottom for entry and exit. Outside this she adds yet other similar envelopes to the number of three or four. Each wrapping is begun at the top as an umbrella and continued roughly parallel with the first.

The actions of the workers are similar to those of the unaided queen. Fresh cells are added round the margin of the existing comb; the innermost wrapping is cut away to make room for extension and fresh coverings are added externally. This process is frequently repeated as the need arises and entails much labour on the workers. The coverings not only afford protection against rain (in arboreal species) but are also of great importance in enclosing layers of air and thus preventing the escape of the heat generated within the nest. Janet and Guiot have found that the temperature within a flourishing nest is about 88° F. and may exceed that of the outside air by 25° F. to 30° F.

When adding fresh "paper" to the existing fabric a wasp works backwards, thus avoiding contact with the newly applied material. This has been reduced to pulp and is then laid on by means of the jaws; all the while the antennæ are kept playing upon both sides of the old and of the new material as though testing its thickness.

In the central portions of the comb the cells are all very regularly hexagonal, but near the margins they are frequently less regular and approach a cylindrical shape. When the first comb has reached a convenient size, which is determined by the dimensions of the cavity in which the nest is being built, a number of pillars of "wasp-paper" are built downwards from the angles of the cells. At the lower ends of these pillars a second tier of cells is built similar to the first, and so on with each successive tier as the needs of the colony demand. Thus the fourth comb is hung from the third, the third from the second, the second from the first, and the whole from the original platform laid by the "queen": this however has ere now been strengthened to bear the increasing load. The distances between the several combs and between these and the innermost wrapping for the time being are just sufficient to allow the inhabitants to move freely about within the nest. The necessary in-

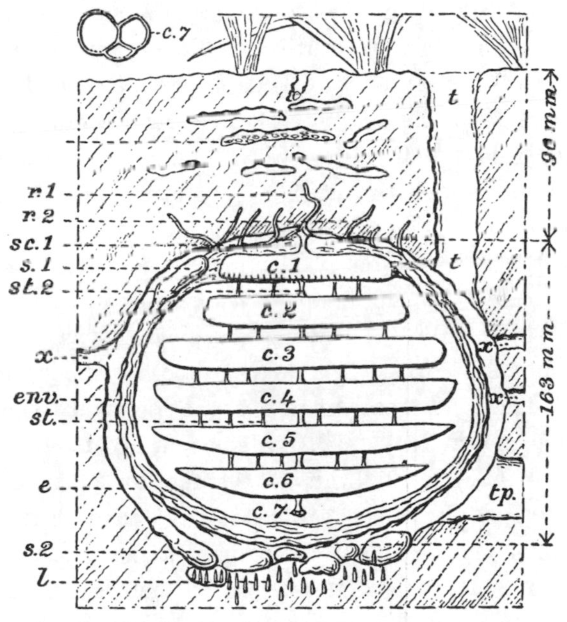

Fig. 28. Diagram of nest of *Vespa germanica*. (After Janet.)

c. 1 to *c.* 7. First to seventh combs of cells. The actual state of *c.* 7 is shown in a separate figure at the top left-hand side. *e.* Space between nest and surrounding soil. *env.* Coverings of nest. *l.* Larvæ of *Pegomyia* (*Acanthiptera*) *inanis*, Fall.? in soil beneath nest. *r.* 1. Root on which nest was originally suspended. *r.* 2. Other roots to which also nest was subsequently attached. *s.* 1. Pebble enclosed in coverings. *s.* 2. Pebbles fallen to bottom of cavity in course of excavations. *sc.* 1. Original stalk for suspension of nest. *st.* 2. First stalk or pillar for suspension of second comb. *st.* Secondary stalks. *t.* Entrance to nest from surface of earth. *tp.* A mole's burrow. *x.* Small lateral passages. A nest of ants (*Lasius flavus*) is indicated in the soil above the nest.

terval is accurately gauged by the workers who may from time to time be seen when building to extend one leg as though measuring their distance from adjacent structures.

The number of combs may be as many as eleven, one below the other; about seven is however the usual number. Their general outline is circular, but they may be of other shapes if the available space demands such departure from the normal. I have seen combs with a distinct rectangular notch upon one side caused by a beam round which the workers had been forced to build for want of room in other directions.

Each cell is not occupied merely once, but so soon as it is vacated by the *imago*, the remains of the dome-shaped bottom of the cocoon are cleared away by the workers, and a fresh egg is deposited by the "queen." The exuviæ and fæcal mass of the previous occupant are left in the cell, still fastened firmly to the roof. There is, however, a limit to the number of times any one cell is employed, probably not more than thrice; the top comb (the oldest) is often to be found with the cells nearly all cut away and reduced to mere stumps; presumably the material removed is used in constructing fresh parts of the nest. It will thus be seen that the number of cells in a nest does not accurately represent the numerical strength of the colony. In a nest containing seven combs Janet counted 11,500 cells, of which over 11,000 had been used twice and about 5000 thrice. A large nest of ten or eleven combs might, then, have had a population of some 50,000 or 60,000.

Towards the end of the summer larger cells are constructed: sometimes a few of these are made in a comb which otherwise consists of ordinary cells; usually the lowest comb contains exclusively these larger "royal" cells. I have a curious comb taken from a nest whose bottom had impinged upon a large stone: it was thus impossible for the wasps to add another tier in this direction: accordingly two patches of "royal" cells were added to one of the existing combs, making two protuberances from its previously circular outline. The "royal" cells are destined for the "queens" of the next generation.

An egg which has been fertilised will in due course produce a female wasp: whether this female be fully developed (a "queen") or but incompletely so (a "worker") depends upon the amount of food with which the larva is supplied. There appears to be no difference in the *quality* of the food supplied to royal larvæ. In fact when the season is favourable and the workers numerous many of the worker larvæ become of greater size in consequence of bountiful supplies of food and become fertile (parthenogenetically). The same phenomenon occurs in nests which have lost their "queen," or in which she has become exhausted. In such cases the workers themselves consume the food which in more favourable circumstances would have been distributed among the larvæ. Such larvæ as there are at the moment of the functional suppression of the "queen" get more bountiful supplies. Hence both these and the younger of the emerged wasps alike become fertile[1]. The males (drones) proceed from eggs which have not been fertilised; these are deposited in ordinary cells

[1] Paul Marchal, *Arch. de Zool. Exp. et Gén.* (3) IV. 1895.

as a rule but occasionally in "royal" cells also. The eggs occasionally laid by large workers invariably produce males, the workers being structurally incapable of impregnation.

With the onset of autumn the strength of the colony rapidly declines. The workers are killed off by the wet and cold; the vigour of the parent "queen" is exhausted and she dies; many of the larvæ are dragged from their cells and devoured by the workers; eventually the whole community, with the exception of the young queens, perishes. Under the attacks of beetle larvæ and other scavengers the fabric of the nest falls into ruin and decay. The survivors have all mated, either within the nest or upon the leaves of plants in the neighbourhood. The males not infrequently mate several times, but are comparatively short-lived. The fertilised queens soon seek safety in the winter quarters described above.

The circumstances which determine whether any given egg be fertilised or not, *i.e.* whether it shall produce in due course a female ("queen" or "worker") or a male, require a few words. It is necessary first to clearly understand the distribution of the sexes among the cells of the combs. Fertilised eggs alone are deposited in the earlier combs: at the end of July or beginning of August fertilised and unfertilised eggs are laid. The latter are not placed in any special cells, but in ordinary worker cells. These cells, however, tend to increase in size and approximate in dimensions to the "royal" cells, being intermediate between them and worker cells. Such intermediate cells may contain large fertile workers, or true queens or males indifferently. Males, however, are here always in far smaller proportion than in the ordinary small cells. Finally the full sized "royal" cells contain invariably (provided the "queen" be healthy) females only, and these as already explained are made into "queens" by liberal diet.

We cannot do better than quote the words of Paul Marchal who has devoted much attention to this matter: "We will admit that after an exclusive and uninterrupted deposit of worker-eggs at first, ..., the reflex, which causes the contraction of the seminal receptacle when each egg is being laid, no longer acts with the same regularity. The eggs can then be laid without being fertilised. Hence the almost sudden appearance of the males depends upon the relative inactivity of the seminal receptacle. It is then that the workers build the large cells and thus give the 'queen' a choice between two different kinds of cell. The large cells, at the end of the season, have the property of stimulating the 'queen,' while she in certain instances seems to visit them with marked preference. We may admit that on these large cells she will concentrate all her energy and that in consequence she will only lay in them fertilised (female) eggs; or again, that she will only lay in them when her seminal receptacle is ready to contract. On the other hand, when she is on the small cells she will lay carelessly and at random, whatever be the state of her receptacle: then according as the receptacle responds or not, the deposit of eggs will produce, in the former case, groups of females, and in the latter, groups of males[1]."

[1] Paul Marchal, *Arch. de Zool. Exp. et Gén.* (3) IV. 1896.

A most instructive observation by the same writer, and one fully appreciated by him, concerns the distribution of sexes in "mixed," (*i.e.* containing both small and large cells) combs. In such a comb "the small cells seem influenced by the proximity of the large cells and contain only a very small proportion of males, ... When the 'queen' is on the large cells her receptacle is in an active condition, and if now and then she gets on to the small cells of the same comb she will have her receptacle in the condition, suited to the large cells and will lay fertilised eggs for most of the time." This explanation satisfactorily removes the will of the "queen" as the determining factor in withholding or bestowing the fertilising element from any given egg.

It is a remarkable circumstance that among honey-bees the case is different. There, it is in the large cells that the unfertilised (drone-producing) eggs are deposited. Drory, however, has shown that if at the time of depositing male-eggs the "queen" bee has no large cells provided for her, then she will place these eggs in ordinary cells: and conversely if only large cells are provided at the time when she is laying fertilised eggs the "queen" bee will nevertheless lay female eggs in the cells originally intended to hold males.

Homing. The power possessed by wasps and other Hymenoptera of finding their way home to the nest has attracted the attention of many naturalists. The observations made by Fabre on several species seem to make it clear that in some cases at any rate it is not by the sense of sight nor by memory that the return journey is made.

On the other hand there is evidence that in other cases the sense of sight is of prime importance. In 1893, a great "wasp year," I found a nest of *V. sylvestris* in an old meat-tin. In order to watch it more conveniently I moved it a few feet: the wasps within the nest, when they came to the exit, paused before taking flight, then they flew to and fro repeatedly over the nest, gradually enlarging the range of flight until they reached the spot where the nest had previously been lying: here all was familiar and they then went straight away. Of those who were abroad from the nest at the time of removal every one returned to the old position and hovered restlessly over it; very few, if any, of these discovered the nest in its new position. After an interval, wasps began returning direct to the nest where I had placed it, without going to the old site at all: these were almost certainly individuals who had been within the nest at the time of removal and had taken its bearings afresh before starting forth on their journeys.

I have observed precisely similar behaviour in bumble-bees kept in captivity. A very slight displacement of the box containing the nest was invariably followed by a careful survey of the surroundings. The bumble-bees on leaving the recently moved box always hovered to and fro over it, facing it, before flying out of the open window.

Again the fossor, *Ammophila hirsuta,* after carefully concealing her burrow with a stone, loose sand and dead grass roots, takes accurate note of its position and constantly assures herself of her ability to return to it from objects in the immediate vicinity.

Others have recorded that the removal of stones, leaves and other small objects from the immediate neighbourhood of the nests of some Hymenoptera has caused apparent confusion and perplexity to the returning insect[1].

It is well known that the sense of smell is developed to a remarkable degree in some insects, nor is it improbable that more senses than one are concerned in the perfection of the "homing" faculty. There is, however, much need of further accurate observation and experiment on these and other animals with like powers.

Food. The food of wasps is of a very varied nature. Ripe fruit, sugary solutions and compounds are familiar to all as their favourite articles of diet. They are also skilful in securing the nectar from many flowers. I find "queens" in early summer freely visiting the flowers of the white *Centaurea* and procuring the nectar in the legitimate manner: later in the year I have known the workers artfully to bite holes at the base of the spur of *Pentstemon* blossoms and steal the nectar without entering the path by which the flower would secure fertilisation. Other observers have recorded blossoms of *Fuchsia* and of *Gladiolus* as being damaged, and various Umbelliferous, Labiate and Scrophularine flowers as being visited by wasps. The gummy substance that is found on the buds and leaves of some plants is also collected and apparently used as a cement in the construction of the nest[2].

[1] Cf. G. W. and E. G. Peckham, *Wisconsin Geol. and Nat. Hist. Surv.* Bull. 2, Sec. Ser. i.; Fabre, *Insect Life* (Trans. D. Sharp).

[2] Janet, *Mém. Soc. Zool. de France*, VIII. 1895.

Wasps are, however, very useful both as scavengers and as destroyers of insect pests. The dead bodies of small mammals, such as mice, are in a few days stript of all flesh by their powerful jaws and rendered inoffensive. I once witnessed a wasp swoop upon a large garden spider that was sitting in the centre of its web and carry it off without in any way disturbing the silken cords of the web. Earwigs, blue-bottle flies and other Diptera, *Aphides* (greenfly and blackfly), small caterpillars and many other insects are captured by them and carried off to feed their larvæ in their combs. It is thus evident that though at times wasps do injury in orchards and gardens, yet they are of considerable service to us in reducing the number of injurious insects.

Parasites. A large number of parasites have been observed in and about the nests of wasps. It is doubtful if the majority of these are really to be regarded as true parasites. More probably they play rather the part of scavengers and are therefore in commensal association with their hosts. The wasps themselves are scrupulously clean in their habits. They always proceed to the entrance at the lowest part of the nest to void their excreta. Hanging on to the edge of the envelopes of the nest they extend the abdomen vertically downwards and discharge their evacuations upon the ground below. Hence beneath the nest there is a considerable accumulation of organic matter. In the soil beneath the nest are usually numerous larvæ of *Pegomyia* (*Acanthiptera*) *inanis*. These Dipterous grubs appear to feed upon the *débris* of food and excrement that proceed from the nest. Their eggs are laid by the fly upon the outer wrappings of the nest and may often be seen as small white specks upon the surface. When the grubs hatch they fall to the bottom of the cavity and bury themselves in the loose soil, damp with the matter that falls upon it. I have however seen the eggs deposited direct upon the mass of filth below the nest. In this case the nest was of *V. sylvestris* and had been built in a library book-case, being suspended from a shelf. The excreta were consequently fully exposed upon the top of the shelf next below. The grubs of *Volucella inanis* have been seen by Janet moving about the combs and visiting cell after cell, apparently devouring the excreta of the wasp larvæ and doing no injury to the occupants. It is remarkable that the adult fly *V. inanis* very closely mimics wasps in its coloration. It has been suggested that this resemblance enables the fly to enter the nest unobserved by the wasps themselves. In this connexion it is worth remembering that within the nest there is very little if any light and that *colour* can be of little, if any, value under such circumstances. Marchal's explanation of the immunity of *Volucella zonaria*, another mimic of like habits, is more probably correct. He observed that the fly moved about within the nest quietly and without excitement and thus avoided attracting attention. On the other hand a strange wasp introduced from one nest into another was greatly agitated, and in spite of similarity of appearance was speedily slain and ejected by the lawful tenants. It is interesting to learn from Janet that a newly hatched wasp when taken to another nest is not thus maltreated.

Among other animals found in the nest may be mentioned a species of *Tachina*, several species of *Acari*, and beetles such as *Dermestes lardarius*, *Bruchus fur* and *Rhipiphorus* (*Metœcus*) *paradoxus*, several species of Microlepidoptera, and an "Ichneumonid" (Janet and Kristof). An undoubted parasite was found by Kristof in the shape of a *Gordius* (Nematode) protruding from the abdomen of a drone. Newstead[1] has found in the nests of *V. germanica* and *V. vulgaris* the Crustacean *Porcellio scaber*; the Acarids *Uropoda elongata* (attached to the Dipteron *Homalomyia canicularis*), *Glyciphaga spinipes* and species of *Tyroglyphus*; the Coleoptera *Leistus rufescens*, *Pterostichus vulgaris*, *Bradycellus verbasci*, *Choleva tristis*, *Homalota succicola*, *Thyamis lurida*, *Quedius puncticollis*, *Epurœa obsoleta*, *Cryptophagus pubescens*, *C. setulosus*, *Rhipiphorus paradoxus*; the Hymenoptera *Aspilota concinna* and species of *Proctotrupes* (?); the Diptera *Cyrtoneura stabulans*, *Homalomyia canicularis*, *H. vesparea*, *Phora rufipes*, *Acanthiptera* (*Volucella*) *inanis*, *Volucella bombylans*, var. *plumosa* and several unidentified species of Lepidoptera. Chitty[2] further records the presence of *Lutheidius minutus* (Coleopteron), and Matthews'[3] *Calliphora erythrocephala* (Dipteron). Many of these animals are merely chance visitors in search of food, or general scavengers.

Of the true parasites the most interesting is the beetle *Rhipiphorus* (*Metœcus*) *paradoxus*. Thanks to

[1] *Ent. Month. Mag.* (2) XXVII. 1891.
[2] *Ibid.* 1893. [3] *Ibid.* 1891.

the researches of Dr T. A. Chapman[1] the life history of this form is now fairly well known. It is almost certain, though not actually proved, that the eggs are laid in cavities in dead wood, posts, palings and so forth, during the autumn. Active hexapodous larvæ emerge from the eggs next spring and by some means find their way into the nests of wasps. On the analogy of what is known to occur in the case of the oil-beetle *Meloë*, it is highly probable that the young larvæ attach themselves to wasps while these are engaged in gathering "wasp-paper" from the surface of the timber, and are thus conveyed within the nest. The larva is black, about 0·5 mm. long, and it possesses a sucker upon the end of the abdomen. By means of this organ it is able to stand erect and paw the air with its feet as though groping for some object. Arrived at its destination the larva eats its way into a wasp-grub and, feeding on the tissues of its host, attains a length of 4·5 mm. It now emerges through the ventral surface of the fourth segment of the wasp-grub and casts a skin which plugs the wound so caused. By this refinement its victim is saved from immediate death and reserved for the full benefit of the *Rhipiphorus*. The larva next seizes the wasp-grub by the second segment and becomes an external parasite upon it. When about 6 mm. long it casts another skin which remains between it and the wasp-grub, the two creatures being now *vis-à-vis*. Eventually the wasp-grub is entirely consumed, but not before it has spun its cocoon within which the *Rhipiphorus* completes its own metamorphosis.

[1] *Ann. Mag. Nat. Hist.* 1870 and *Ent. Month. Mag.* (2) XXVII. 1891.

It emerges as a perfect beetle a day or two after the wasps in the surrounding cells. The cells of infected specimens at first appear whiter than the rest owing to the glistening white *Rhipiphorus* larva shining through the silk cap. Later, when emergence is approaching, they appear blackish or reddish as opposed to the greenish tint of those occupied by wasps. From the relative positions of the host and parasite it will be evident that whereas a wasp-grub always faces towards the centre of the comb, the beetle-grub faces towards the circumference.

The species of British Wasps (Vespidæ). It has already been pointed out in what particulars the true social wasps differ from their immediate allies and other insects with which they are at all likely to be confused. The genital armature of the drones furnishes useful features in the determination of species. This apparatus consists of two pairs of forceps. The outer, or *stipites,* spring from a basal ring, the *cardo*; the inner lie between the stipites and are called the *sagittæ.* Below are given the diagnostic characters of our seven British species.

1. *Vespa crabro,* the Hornet: thorax reddish brown. May be known by its great size, though "queen" wasps are often mistaken for Hornets: the colour of the thorax is a sure guide. Nests in hollow trees, outhouses or less frequently in banks, in which case the nest is always very near the surface.
2. *Vespa vulgaris*: thorax black and yellow; face short, *i.e.* no, or very little, interval between eyes and base of mandibles; abdomen yellow and black, markings sharply defined; lateral yellow stripe on pronotum straight- and parallel-sided. ♂ decidedly hairy. 1st joint (scape) of antenna of ♂ with

yellow spot in front, of ♀ and ☿ black. Sagittæ of male genital armature not hollowed out at apex, but simply rounded. Nests underground. Very common.

3. *Vespa germanica*: colour of thorax, shape of face, and colour of abdomen as in *V. vulgaris*, but lateral yellow stripe on pronotum decidedly convex towards ventral surface. ♂ slightly hairy. Antennæ as in *V. vulgaris*. Sagittæ of male genital armature distinctly hollowed out at apex. Nests underground or in thatch etc. Very common.

These, our two commonest species, are very much alike. There are certain differences other than those given above, but they are not absolutely constant; the subjoined table of comparison will assist in distinguishing the two.

	V. vulgaris	*V. germanica*
Longitudinal yellow stripes on side of pronotum	Straight- and parallel-sided	Enlarged ventrally
Yellow blotch in concavity of eyes	Hollowed out	Not hollowed out
Black marks on clypeus	Median black line often dilated at lower end; sometimes interrupted or reduced to a spot, but not flanked by two other black spots	Median black line flanked at lower end by two black spots; the line often reduced to a mere spot
Tibiæ	Often spotted with black	Usually entirely yellow

4. *Vespa austriaca*: colour of thorax, shape of face and colour of abdomen as in (2) and (3); scape of antenna yellow in front in ♂ and ♀: abdomen glossy, carrying thick black hairs; tibiæ with long projecting hairs. Rare: ☿ not known. Genital armature of ♂ narrower, paler, and its sides more parallel than in *rufa*.

5. *Vespa rufa*: colour of thorax and shape of face as in (2), (3) and (4), but black markings on abdomen are blurred with brown and not sharply defined. Abdomen glossy. Nests underground. Not uncommon. Armature more robust and darker.

   [Carpenter and Pack-Beresford[1] appear to have satisfactorily established that *V. rufa* and *V. austriaca* are but dimorphic forms of the same species. The latter they regard as the ancestral type, the former as a more recent development. In England the younger and more vigorous *V. rufa* has almost completely displaced *V. austriaca*, which however still holds its own in Ireland.]

6. *Vespa sylvestris*: colour of thorax and of abdomen as in (2), (3) and (4), but face long, *i.e.* a well-marked interval between the eyes and base of mandibles. Clypeus with only a small central black spot. Scape of antennæ yellow in front in ♂, ♀ and ☿. Tibiæ with long projecting hairs. Stipites of armature with inner margin deeply sinuate near the base, then parallel and almost in contact to the apex, where they terminate in a sharp spine without any fringe of hairs. Nests on branches of trees, bushes, etc.; occasionally underground, near surface, or in old tins, pots, etc. Common.

7. *Vespa norwegica*: colour of thorax as in (2) etc.; shape of face, scape of antennæ and tibiæ as in (6). Clypeus with broad central black line often dilated in middle. Abdomen generally reddish at base. Stipites of armature with inner margin slightly sinuate near the middle, and far apart throughout their length; their apex furnished with a spine and a dense fringe of hairs beyond it. Nests on branches of bushes and shrubs, occasionally on trees. Fairly common.

[The males of each species can always be distinguished from females, whether "queens" or "workers," by the following:—antennæ of males have 13 joints; females only 12 joints; the abdomens of males have seven visible segments, those of females only six; the wings of males are longer and relatively narrower than those of females; and lastly, males cannot sting.]

[1] *Irish Naturalist*, XII. 9. Sept. 1903.

CHAPTER VI.

THE FRESH-WATER MUSSEL.

LAMELLIBRANCH Molluscs of the Family *Unionidæ* are to be found in nearly all ponds and streams. They are, in spite of their large size, frequently overlooked in consequence of their habit of lying almost totally buried in the mud at the bottom of the water. Their presence is usually betrayed by broken shells lying upon the banks or exposing the glittering inner surface beneath the water. These fragments are the relics of the repasts of swans, coots and other water-fowl or occasionally of water-voles.

The shell of *Anodonta cygnea*, the Swan Mussel, perhaps the commonest of our *Unionidæ*, is from four to six inches in length, from two to three inches broad, and from one to two and a half thick from right to left. It is of a greenish-brown tint, but is often speckled with white patches in consequence of the surface layer being corroded and the deeper calcareous layer exposed. The two valves are placed on the right and left of the animal's body and resemble each other in shape (equivalve). Each is a reflex (or looking-glass-picture) of the other. The hinge-line, along which the two valves are united by a strong elastic ligament, is straight and corresponds with the dorsal surface of the animal. The outline of the rest of the shell may be described as roughly oval, the long curved margin being ventral, the blunter end anterior and the more tapering posterior.

The external surface is marked by concentric lines parallel with the margin; these "lines-of-growth" are in sets which probably each represent the growth of one year. Very few exact observations have been made as to the rate of growth, nor are the artificial conditions of an aquarium suitable for experiment. Buxbaum[1] has however recorded a valuable instance: a specimen of *Anodonta cellensis*, known to be not more than three years old, measured 135 mm. in length, 75 mm. in breadth and 45 mm. in thickness; its shell exhibited two strongly marked lines marking out three sets of more faintly marked areas which may reasonably be supposed each to represent the increase gained in a single year. The umbo is the oldest part of the shell and the line nearest to it (usually very faint) marks the margin of the original shell of the young mussel. The curvature of the lines of growth is not quite smooth, but at about the middle of the ventral surface tends to become convex in the dorsal direction: this irregularity, the cause of which will be mentioned later (*vide* p. 193), is more marked in the earlier lines of growth than the later, and is more evident in some specimens than in others; it is usually pronounced in species of *Unio* and less conspicuous in *Anodonta*. The

[1] *Zoolog. Garten*, XXXI. 1890.

elastic ligament posterior to the "umbones" serves to open the shell ventrally by pulling the dorsal margins of the valves together; its action is opposed by the contraction of the two adductor muscles. The "umbones" serve as fulcra upon which the valves move as levers. Hence in a dead specimen the valves always gape ventrally.

The shell is composed of three layers; an outer, organic, coloured periostracum, composed of "conchiolin" secreted by the margin only of the mantle; a middle, calcified, white prismatic layer, forming in *Anodonta* about half the thickness and also secreted by the mantle margin; and an inner lamellated calcareous mother-of-pearl or nacreous layer. All round the margin of the shell the periostracum projects and is slightly reflected as a flexible sheet beyond the other layers. The nacreous layer is iridescent owing to the optical effect produced by the edge of the alternating lamellæ of conchiolin and of conchiolin charged with calcareous matter. This layer is secreted by the entire surface of the mantle and throughout life; hence it steadily increases in thickness and is thickest in the oldest portions of the shell. Pearls, which are often found in *Anodonta*, are formed by this material (*vide infra* p. 198). The ligament consists of an outer layer continuous with the periostracum and an inner which is striated radially; the latter is continuous with the nacreous layer of the shell and at its margin the radial fibres are connected with numerous muscle fibres.

In *Anodonta*, as the generic name implies, the valves are destitute of hinge-teeth. In such lamellibranchs as possess the full complement of hinge-teeth those placed beneath the umbones are termed "cardinal-teeth," in front of these are the "anterior lateral-teeth," and behind are the "posterior lateral." *Unio*, while possessing both anterior and posterior laterals, has no cardinal-teeth

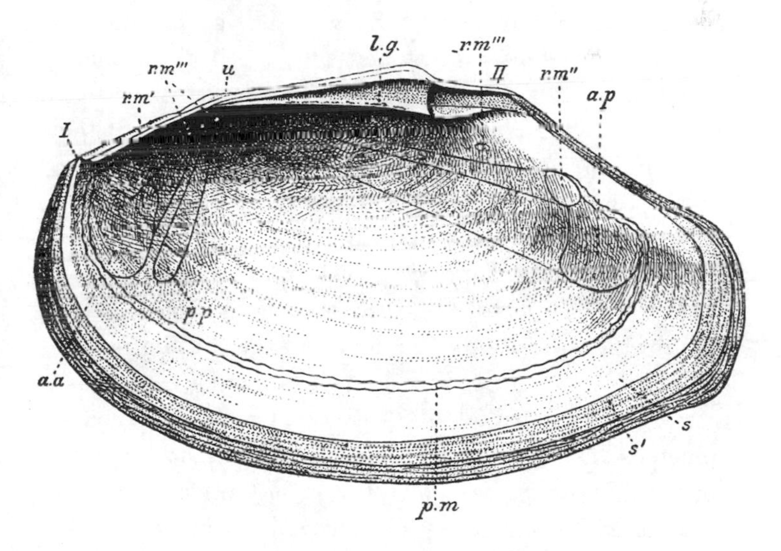

Fig. 29. The right valve of *Anodonta cygnea* from within. (After Howes.)

If carefully dissected from the body under water, it will be seen that the superficial chitinous layer is continuous on the dorsal side, between the points I. II., and also for the area marked *s'*, where it is turned in and reflected on to the pallial muscle. The lines drawn in the interior of the valve indicate impressions which mark the points of origin, the shifting during growth and the final attachment of the muscles.

a.a. Anterior adductor muscle. *a.p.* Posterior adductor muscle. *lg.* Ligament. *p.m.* Pallial muscle. *p.p.* "Protractor pedis" muscle (retractor of shell). *r.m'.* Anterior "retractor" muscle (protractor of shell). *r.m''.* Posterior "retractor" muscle (protractor of shell). *r.m'''.* Lesser "retractor" muscle (depressor of shell). *s.* Valve of the shell. *s'.* Inturned edge of shell. *u.* Umbo.

The inner surface of each valve is marked by certain muscular impressions. In the anterior region is a large shallow pit due to the attachment of the anterior adductor muscle, one of those which close the shell, and of the (so called) anterior retractor muscle of the foot. Posterior to this impression and about level with its ventral end is a smaller pit to which is fastened the (so called) protractor muscle of the foot. In the posterior portion of the shell is the large posterior adductor impression, and immediately dorsal and in front of it the smaller impression of the (so called) posterior retractor of the foot. Diverging from the region of the umbo towards the final position of these several chief muscles are faint tracks which mark the lines along which the muscular attachments have shifted during the growth of the shell. The details of the shifting process are not known. Parallel to the ventral border of the shell there runs the "pallial line" caused by the insertion of numerous muscle fibres lying in the mantle (pallial muscles): the line extends from the anterior to the posterior adductor impressions. In addition to these well-defined markings there are others less easy to detect arranged in two groups, the one near the anterior and the other near the posterior limit of the ligament. To these small impressions are attached groups of minor retractor (or better, *depressor*) muscles. The universally accepted terms of "retractor, or protractor, pedis" as applied to the above-mentioned muscles are unfortunate. I have no doubt that, so far from drawing the foot backward or forward, their real function is to drag the shell and therefore the whole animal forward or backward, the foot being fixed and the shell relatively movable (*vide infra, sub* Locomotion, p. 171).

The mantle, to which reference has already been made, is a fold of the dorsal surface of the animal hanging down on the right and left sides, and conforming in extent and outline with the valves of the shell. The thickened edges of the folds are not united except opposite the posterior adductor muscle, where they are fused for about half-an-inch. Just below this region of fusion each edge is pigmented, and withdrawn twice in quick succession from its fellow. Two slits are thus left between the right and left mantle lobes at the posterior end of the animal. The lower slit, whose walls are provided with a fringe of short tentacles, is the inhalant siphon through which currents of water are drawn into the cavity beneath the mantle by the action of cilia, which clothe all the free external surfaces and are especially pronounced upon the gills and labial palps. The water passes at first forwards and dorsally through the lattice-work of the gills, it then is directed posteriorly and eventually leaves the shell by the upper slit, the exhalant siphon.

When at rest and undisturbed in its natural habitat the mussel buries fully the anterior three-fourths of the length of the shell in the soft mud. Indeed it requires some experience to detect their presence, for frequently the only parts visible are the margins of the two widely opened siphons enclosing an **8**-shaped space. The "waist" of the **8** is formed by a small lobe of the mantle of one side overlapping the other side so as, functionally, to separate the exhalant from the inhalant siphon. In the necessarily artificial conditions of an aquarium the animal, in my experience, never buries itself so deeply. The easiest method of securing specimens from situations out of arm's reach is to cut a long stick to a fine point and carefully insert the pointed end into the inhalant siphon; the animal immediately closes the valves of the shell, and with such firm pressure that it clings to the improvised fishing-rod, and may be bodily withdrawn from the water. This method, which is simple enough as soon as the eye has become accustomed to the refraction due to the surface of the water, has the advantage of not stirring up the mud, so that the water remains clear and other specimens are obtainable at the same spot.

The movements of the

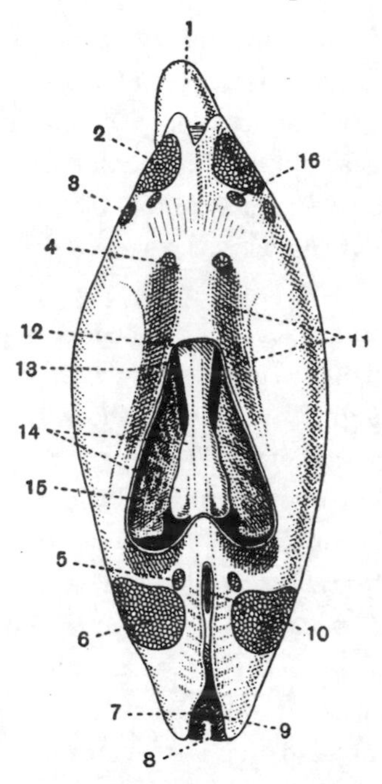

Fig. 30. Dorsal view of *Anodonta mutabilis*, with the upper wall of the pericardium removed to show the heart × about 1. After Hatschek and Cori.

1. Foot. 2. Anterior adductor muscle. 3. Retractor muscle. 4. Anterior protractor muscle. 5. Posterior protractor muscle. 6. Posterior adductor muscle. 7. Dorsal siphon. 8. Ventral siphon. 9. Anus. 10. Split between left and right mantle lobes through which larvæ may leave the epibranchial chamber. 11. Keber's organ. 12. Rectum traversing ventricle. 13. Nephrostome or internal opening of organ of Bojanus. 14. Ventricle. 15. Left auricle. 16. Anterior protractor muscle.

[The muscles are named in accordance with their respective functions.]

animal are very sluggish, and are accomplished in a manner very different from the gliding motion of a gastropod. The rate of progression of *Unio margaritifer* has been recorded by Boycott and Howell as about 15 feet a day, or roughly a mile a year. I have noticed that specimens kept in aquaria do not, as a rule, move during the daytime, but

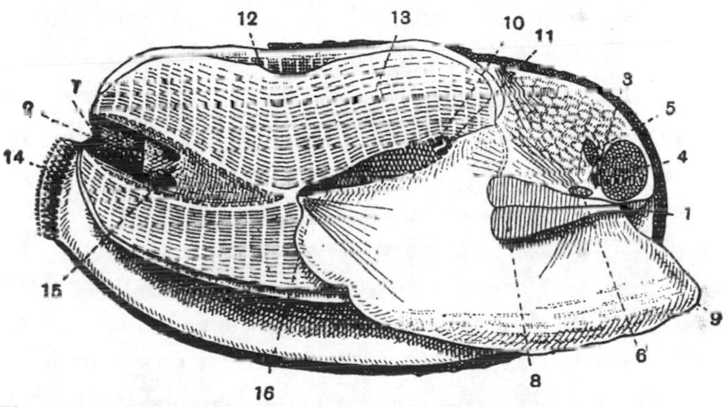

Fig. 31. Right side of *Anodonta mutabilis* with the mantle cut away and the right gills folded back × about 1. From Hatschek and Cori.

1. Mouth. 2. Anus. 3. Cerebro-pleural ganglion. 4. Anterior adductor muscle. 5. Anterior protractor muscle. 6. Retractor muscle. 7. Dorsal siphon. 8. Inner labial palp. 9. Foot. 10. External opening of nephridium or organ of Bojanus. 11. Opening of genital duct. 12. Outer right gill-plate. 13. Inner right gill-plate. 14. Ventral siphon. 15. Epibranchial chamber, the inner lamellæ of the right and left inner gills having been slit apart. 16. Posterior protractor muscle.

become active a few hours after sunset, and further that the activity of one individual almost invariably appears to stimulate its companions to exert themselves, even though they may be situated several inches apart.

The organ of locomotion is the "foot," a strong muscular and bluntly hatchet-shaped projection of the mid-ventral surface. When at rest the foot is concealed within the shell. Its protrusion is effected between the ventral and anterior margins of the valves, whence it may sometimes be seen projecting as a tumid, yellowish-pink, fleshy mass, for a distance of several inches. I have seen the foot reflected over the shell of a large specimen to such an extent as to cover almost the entire valve and to touch the inhalant siphon with its apex. When the animal wishes to move forward the foot gradually extends, insinuating itself into the mud or into narrow crevices between pebbles and stones in such a way as to secure a firm hold by lateral pressure; the sides of the foot, especially in the more ventral and anterior regions, then swell so as to press even more firmly against the enclosing mud, and at the same time by a convulsive heave the shell rides forward on the foot, and then slightly downward. During this movement of the shell very little, if any, of the foot disappears within the shell; in fact the valves are at the moment forcibly drawn together so as to pinch the foot and prevent the shell from slipping down over it, or the foot itself from being drawn back into the shell. Hence the effect, so long as the valves are approximated, is that the shell is pulled evenly forward in a horizontal direction: when the adductors relax and the valves gape apart the shell drops down over the foot for a slight distance. In this way the animal advances step by step; there is no steady forward gliding as in gastropods.

At the close of each progressive movement the valves, as already stated, separate slightly, but the mantle edges are kept firmly applied to one another so that there is a decided negative pressure within the shell cavity. This can be very plainly seen to be the case in the region of the siphons. These remain projecting posteriorly but are closed, and, at the moment of relaxation of the adductors and consequent separation of the valves, are forcibly and suddenly flattened from side to side. It is possible that advantage is taken of this hydrostatic pressure outside the animal in forcing the shell downward into the soft mud. Immediately afterwards the mantle borders separate along their entire length, water rushes in everywhere and the siphons reopen.

Prior to each fresh effort in a forward (or backward) direction the mantle borders are again approximated, nor is any effort made until their closure is complete. These facts were ascertained by repeated observations of a specimen lying partly on its side close against the glass of an aquarium. The forward movements described are brought about by the contractions of the so-called retractor muscles of the foot: it is obvious that "protractors of the shell" would be a more appropriate title. Progression, however, is not the only movement of which the mussel is capable; I have seen specimens move backward out of a narrow cleft after vain endeavours to force a passage, or in order to make a second attempt to get round an awkward corner, and this they can do just as successfully and forcibly as forward locomotion. In this case the shell rides backward over the foot, being drawn back by the so-called protractor of the foot (retractor of the shell). There is considerably more power of surmounting obstacles than might be expected: I have frequently seen large specimens get over a rough stone some three inches high. Normally, however, they prefer to move along the level muddy bottom, in which they plough furrows that may often be tracked for several yards.

The method by which the protrusion of the foot is brought about has been the subject of much discussion[1]. When fully distended this organ becomes far more transparent in appearance and is obviously dilated by some fluid substance within. It has been maintained by some investigators that water-pores exist through which water is taken in for the purpose of dilation. Fleishmann[2] and others have, however, satisfactorily established that the supposed pores are either artificial injuries or the apertures of mucous glands affording no communication between the blood and the outer water. The entire phenomenon is dependent upon the distribution of the blood within the body. During rest the blood, which forms fully half the weight of the body, is largely contained in the sinuses of the mantle. When the foot is required for use[3] an inrush of blood is permitted into the pedal lacunar spaces, but egress is prevented by a special sphincter muscle, known as "Keber's valve," which closes the great vein leading from the foot to the nephridia. The tissues of the foot thus become swollen and turgid and relatively firm for locomotor purposes. It is noteworthy that there is no actual increase in the volume of the soft parts of the animal

[1] Cattie, *Zool. Anz.* VI. 1883. Griesbach, *ibid.* VII. 1884.
[2] *Zeit. wiss. Zool.* XLII. 1885.
[3] Willem, *Mém. Cour. Acad. R. Sci. Belg.* LVII.

during protrusion; such alteration as there is is merely increase of one part balanced by diminution of others. The retraction of the foot after protrusion is not effected by the "retractors" of the foot, but by the removal of the blood aided by the contractions of the intrinsic muscles which form so large a part of the ventral portion of the foot and which are highly stretched during the period of turgescence.

Food and Digestion. The food of mussels is obtained solely by aid of the ciliary currents already referred to. The tentacles surrounding the inhalant siphon exercise a certain power of selection, being extremely sensitive to changes in the quality of the water. On the approach of a substance of disagreeable flavour the adductors are violently contracted and a strong current of water forcibly discharged through both siphons washes the objectionable matter several inches away. A fragment of tobacco ash from my pipe falling through the water close to the inhalant siphon of a specimen under observation was treated in this way.

The water is directed into the mouth along the inverted groove formed by the ciliated labial palps of which there is one pair upon each side. I am disposed to think that nutritive particles are in some way caught in mucous discharges and thus swept into the mouth, for frequently in freshly captured specimens there is a line of food-laden mucus between the palps and an accumulation of the same substance at the mouth itself. The food consists of various microscopic organisms and *débris* of larger creatures. An examination of the contents of the alimentary canal of an *Anodonta cygnea* taken from Busbridge Ponds, near Godalming, and dissected within an hour of capture revealed carapaces and fragments of limbs of minute crustacea, shreds of epidermis of dicotyledonous leaves and of wood fibre, filaments of *Spirogyra*, and enormous numbers of a spirilliform bacterium, as well as the following[1]:—numerous remains of Rotifers, Rhizopods, etc., Sponge Spicules, *Diatomaceæ* of the following species, *Frigilaria capucina* Desm., *Cocconeis placentula*, Ehrenb., *C. pediculus* Ehrenb., *Navicula Brébissonii* Kütz, *N. exilis* (Kütz), Grun., *N. radiosa* Kütz, *Eunotia pectinalis* (Dillw.), Rabenh., *Nitzschia palea* (Kütz), Sm., *N. acicularis* (Kütz), Sm., *Pleurosigma attenuatum* (Kütz), W. Sm., *Diatoma elongatum* Ag., *Cymbella lanceolata* (Ehrenb.), Kirchn., *C. cymbiformis* (Kütz), Bréb., *C. cistula* (Hempr.), Kirchn. var., *maculata* (Kütz), Grun., *Melosira varians* Ag.; five species of *Desmidiaceæ,* viz.: *Cosmarium præmorsum* Bréb., *C. Turpinii* Bréb., *C. botrytis* Menegh., *C. granatum* Bréb., var. *subgranatum* Nordat. and *Staurastrum crenulatum* (Näg.) Delp., and three species of Protococcoideæ, *Scenedesmus quadricauda* (Turp.), Menegh., *Pediastrum Boryanum* (Turp.), Bréb., and *Dictyosphærium Thunbergianum* Näg. It is remarkable that a large number of these Algæ were found to be in a fragmentary condition, probably as the result of digestive changes undergone within the stomach or intestine.

After entering the mouth the food is swept by

[1] **Kindly identified by Prof. G. S. West of the Royal Agricultural College, Cirencester.**

the action of cilia upwards into the stomach and thence along the looped intestine, whose walls are ciliated throughout, to the anus situated above and behind the posterior adductor muscle. Into the cavity of the stomach there discharge the ducts of the large digestive

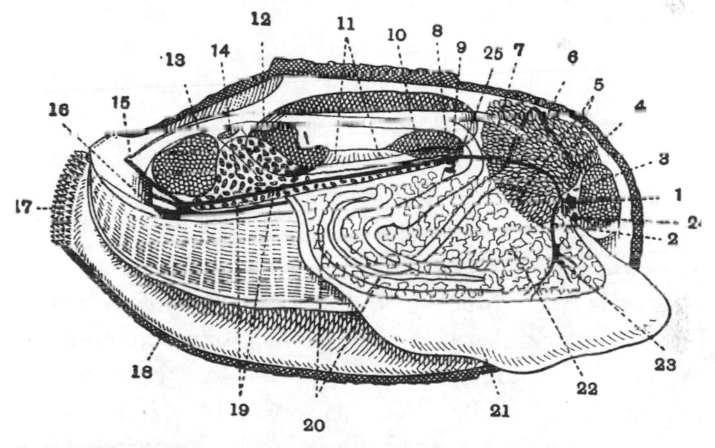

Fig. 32. Right side of *Anodonta mutabilis*, dissected to show the viscera × about 1. From Hatschek and Cori.

1. Cerebro-pleural ganglion. 2. Cerebro-pedal commissure. 3. Œsophagus. 4. Anterior retractor muscle. 5. Liver. 6. Stomach. 7. Aorta. 8. External opening of organ of Bojanus or nephridium. 9. Nephrostome or internal opening of the same. 10. Pericardium. 11. Right auricle. 12. Posterior end of ventricle passing into posterior aorta. 13. Rectum. 14. Glandular part of nephridium. 15. Anus. 16. Opening of epibranchial chamber. 17. Ventral siphon. 18. Edge of shell. 19. Cerebro-visceral commissure. 20. Intestine. 21. Foot. 22. Reproductive organs. 23. Pedal ganglion of right side. 24. Mouth. 25. Opening of the reproductive organ.

gland ("liver"), whose secretion has been shown by Griffiths to effect the conversion of starch into dextrose sugar, proteids to peptones, and to emulsify fats. It is highly probable that this gland is also an important organ for the absorption of digested products. The walls of the stomach are thin and produce a cuticular lining, the *flèche tricuspide,* which is thought by some authorities to serve as a protection to the secretory cells against hard or sharp food particles.

The posterior region of the stomach gives off upon its right side a cæcal diverticulum, which, however, is not completely separated either from the stomach itself or from the intestine, but appears rather as a fold of the latter. By constriction of the lips of the fold communication between the cavity of the cæcum and those of the stomach and intestine can probably be closed. In *Dreissensia polymorpha,* another British fresh-water lamellibranch, the cæcum is entirely separated from the intestine as a distinct tube opening into the stomach. Within the cæcum is developed a remarkable structure, the crystalline style. This is a rod-like body, some two or three inches in length, tapering posteriorly, and composed of rather firm transparent gelatinous material, arranged in concentric layers. It is probably a special development of the *flèche tricuspide.* The precise function of the crystalline style has been variously assigned. By some it is considered to act as a pestle and to serve in mixing the food with the gastric ferments, the style itself becoming at the same time softened and dissolved, furnishing a soft glutinous envelope to any angular particles in the food (Schultze). According to Hazay[1], from spring to autumn the stomach is filled with a

[1] *Malacozool. Blät.* III. 1881; cf. Krukenberg, *Vergl. physiol. Vorträge,* II. 1882, Heidelberg.

glutinous mass in which the crystalline style is slowly differentiated, the process being complete by October. The remaining jelly, apparently superfluous food material, passes into the first section of the intestine and by November the stomach is empty of it. In the intestine the jelly becomes compacted to form the crystalline style which is slowly consumed during the winter months. Haseloff[1] maintains that the style is a reserve of food, disappearing during periods of fast and reappearing when food is taken. Mitra[2], who has recently investigated the matter, adopts an entirely different explanation; in his opinion the style is a digestive ferment whose function is to convert starch to sugar; he finds that much food surrounds and is embedded in that end of the style which projects into the stomach and that the style itself is formed and vanishes again about once a day, being present whenever digestion is going on; it is a colloid proteid of the globulin class, is soluble in water and is probably excreted by the "liver" as a viscous fluid.

I am unable to confirm Mitra's results; specimens that had been fasting for two days (in October) I found to contain well developed styles, these after removal from the body remaining undissolved in cold water for three days and were then removed; when fresh styles were placed in dilute solutions of starch for several days I was unable to detect the presence of sugar; the proteid reactions of the style were however very well marked and there were indications of the presence of peptone. It is evident that

[1] *Biol. Centralbl.* VII. and *Kiel. Inaug. Diss. Ostarode*, 1888.
[2] *Q.J.M.S.* XLIV. 1901, with literature references.

a long and careful series of experiments and observations is still needed here; they should be conducted upon specimens taken fresh from their native waters and not upon individuals kept in aquaria.

The food appears to pass very slowly along the intestine for it may be found there after specimens have been kept for several days in perfectly clear tap-water. The ventral wall of the last section of the intestine, or rectum, where it runs dorsally from the visceral portion of the foot towards the pericardial cavity, is folded so as to form a prominent ridge, the typhlosole, projecting into the gut cavity. This fold increases the internal absorbing surface and narrows the channel along which the food is passing, thereby possibly accentuating the effect of the cilia upon the fluid contents. The fæces are often expelled with considerable violence, being shot forth a distance of six or eight inches. They appear to me to travel in a small body of water moving as a vortex ring, for though their own movement is rapid there is no apparent disturbance of the surrounding water which they traverse. I am unable to state by what muscles this forcible defæcation is produced: no movement of the shell-valves could be detected.

Circulation, Respiration[1]. The heart, which consists of a single ventricle wrapped round the rectum and a pair of triangular auricles, is situated dorsally close beneath the mantle. It lies in the pericardial space which it practically fills during life. The pulsations can be plainly

[1] Langer, *Denkschr. Wien. Akad.* VIII. Abth. 2; Willem. *op. cit.*; Owsjannikow, *Bull. Acad. Imp. Sci. St Petersb.* (5) II.

seen through the thin overlying structures after removal of the animal from the shell; their rate is from four to six per minute but it varies with the temperature and with the intraventricular blood-pressure. Two main arteries, one from each end of the ventricle, convey oxygenated blood from the heart to all parts of the body. The anterior aorta runs dorsal and the posterior ventral to the rectum. After traversing the lacunar spaces of the foot and viscera the blood is returned to a large median *vena cava* on the ventral side of the pericardium; hence it flows along the two afferent branchial vessels through the nephridia to the gills, where it receives fresh supplies of oxygen and parts with its carbon dioxide; the efferent branchials return the blood from the gills to the auricles, receiving also oxygenated blood returning from the mantle lobes. The last named portion of the blood does not pass through the gills at all; it is thus evident that the mantle lobes are of equal importance with the gills as respiratory organs. The auricles drive the blood into the ventricle through valved apertures by which reflux is prevented. The pressure within the auricles during their relaxation is negative and thus blood is, as it were, sucked back towards the heart. Valvular arrangements are present in the form of sphincter muscles which can constrict the cavity of the vessels in whose walls they are placed. Of these the most important is "Keber's valve" already mentioned in connexion with the turgescence of the foot.

The blood is colourless, and contains colourless amœboid corpuscles which take up and absorb degenerating tissue and foreign bodies which have entered the system. The plasma contains large quantities of calcareous salts in solution. The respiratory interchange of gases, though primarily taking place in the gills and

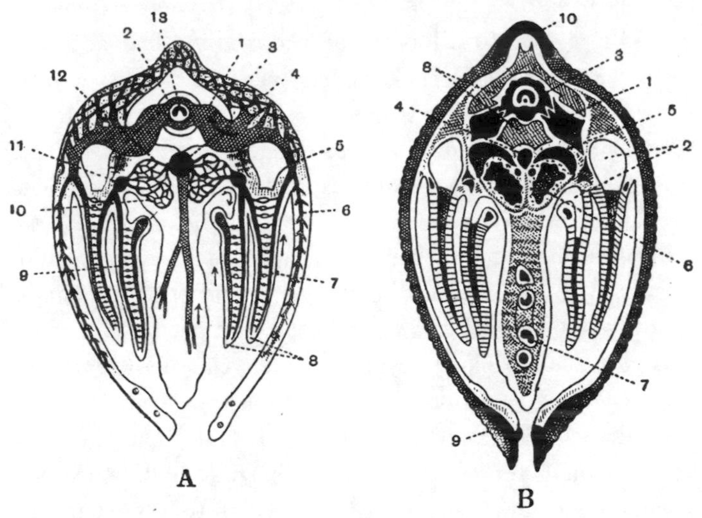

Fig. 33. A. Diagrammatic section through *Anodonta* to show the circulation of the blood. B. Section through *Anodonta* near the posterior edge of the foot. From Howes.

A. 1. Right auricle. 2. Ventricle. 3. Keber's organ. 4. Vena cava. 5. Efferent branchial trunk. 6. Efferent pallial vessel. 7. Efferent branchial vessel. 8. Branchiæ. 9. Afferent branchial vessel. 10. Efferent renal vessel. 11. Afferent branchial trunk. 12. Afferent renal vessel. 13. Rectum.

B. 1. Right auricle. 2. Epibranchial chamber. 3. Ventricle. 4. Vena cava. 5. Non-glandular part of the kidney. 6. Glandular part of the kidney. 7. Intestine in foot. 8. Pericardium. 9. Shell. 10. Ligament of shell.

mantle, is also very largely effected by the surface of the foot, especially when this is distended, and by that of the visceral region of the body.

The orange-coloured pigment, which is universally present in the skin and particularly brilliant in that of the foot, and also on the surface of the nerve ganglia, is known as tetronerythrin and possesses a strong affinity for oxygen. It is noteworthy that this respiratory pigment is especially well developed in the foot, where, as already indicated, the blood is in a stagnant condition whenever the animal is moving. There can be little doubt that the blood imprisoned in the pedal lacunæ during turgescence is maintained duly supplied with oxygen by the united action of this pigment and diffusion accelerated by the stretched condition of the skin.

Excretion[1]. The elimination of waste products is brought about by the portal circulation within the glandular substance of the nephridia or organs of Bojanus[2], as they are often termed. These organs lie right and left ventral to the pericardium, with whose cavity they communicate, thus establishing connexion between it and the outside water through the ureter. Within their substance numerous nitrogenous compounds allied to urea and uric acid have been identified.

Keber's organs, a pair of rusty red glandular proliferations of the anterior walls of the pericardium, also perform similar excretory functions. They are supplied with

[1] Griesbach, *Arch. f. Naturges.* XLIII. 1877; Krukenberg, *Vergl. physiol. Stud.* I. 2; Rankin, *Jena Zeitsch.* XXIV. 1889–90.

[2] Louis Henri Bojanus, born July 16, 1776, at Buschweiler in Alsace; died April 2, 1827, at Darmstadt; Professor at the University of Vilna. He was especially noted for his works on comparative anatomy and veterinary medicine.

arterial blood returning from the mantle and external gills: their excretion is probably discharged into the pericardium and thence passes out through the reno-pericardial aperture and ureter. On reaching the surface of the body the excreted matter is caught in the out-flowing current of water and ultimately discharged at the exhalant siphon.

Nervous System and Sense Organs. The nervous system consists of three pairs of ganglia, connected by cords and innervating the regions in their neighbourhood. The most anterior ganglia (cerebrals, or according to some, cerebro-pleurals) are placed one on each side of the mouth just ventral and anterior to the retractor muscle of the shell; the two are united dorsally by a commissure which runs round the front of the mouth, and are each connected with the other ganglia of the same side of the body, viz. the pedal and the visceral. The two pedal ganglia are close together in the central part of the foot at the junction of the visceral and muscular portions. The visceral ganglia are situated upon the ventral surface of the posterior adductor muscle and are so closely opposed as to form, with their nerves, a single conspicuous stellate mass. The anterior ganglia supply nerves to the labial palps, the anterior adductor muscle and neighbouring structures, and also give off each a nerve which runs with the cerebro-pedal connective to the otocyst. Each pedal ganglion gives nerves to the foot and appears, for the reason just mentioned, to supply the otocyst. From the visceral ganglion mass nerves are distributed to the posterior adductor, the mantle, siphons and other posterior portions.

Occasionally, according to Hartog[1], there is to be found on one or both of the cerebro-visceral connectives a minute ganglion, situated in front of the pericardium at the point where the right and left commissures diverge from one another. This observation I have recently been able to confirm[2].

The organs of special sense are few and inconspicuous. The tentacles surrounding the inhalant siphon are undoubtedly tactile and gustatory in function. They are also keenly sensitive to changes of light and shade, and this in spite of absence of all special optic nerve-endings. Other portions of the body are also sensitive to light, particularly the surface of the foot, as I have proved by repeated experiments in which the siphons were screened from the source of light[3]. Tactile cells, or nerve-endings, whose outer free ends bear fine setæ, are abundant on the edges of the mantle, especially near the siphons, and less plentifully upon the labial palps, the inner surface of the mantle and on the foot. An olfactory epithelium, or osphradium, is present covering the ventral surface of the visceral ganglia and extending a short distance along a pair of stout nerves which proceed thence to the gills and contain numerous ganglion cells. The supposed auditory organs, the otocysts, lie a little way behind the pedal ganglia. Each is a small vesicle whose walls are composed of sensory and of ciliated cells; the cavity of the vesicle is occupied by fluid in which is a single spherical calcareous

[1] Howes, *Atlas of Pract. Elem. Zool.* 1902. Expl. of Fig. XIX. Pl. XXII.

[2] Latter, *Nature*, LXVIII. Oct. 1903, p. 623.

[3] Cf. White, *Ann. Mag. Nat. Hist.* 1869.

otolith. There is no evidence that Anodonta is sensitive to sound as we understand it[1]: it is far more probable that the otocysts are equilibrating organs. Possibly they are concerned in detecting vibrations of the water or of the pond-bottom. The animals are certainly extremely sensitive to vibrations when in captivity: unless the aquarium or dish in which they are kept be placed upon a table which is not affected by vibrations of the floor, *e.g.* a slate balance-table affixed to the walls direct, it is extremely difficult to approach them without causing every specimen to retract its foot and close the shell.

Reproduction and Development[2]. In all the *Unionidæ* the sexes are normally separate, though occasional instances of hermaphrodite *Anodonta* have been found. It is often stated[3] that the shell of the female *Anodonta* or *Unio* can be distinguished from that of the male by its greater convexity and increased transverse diameter. As a matter of fact there is no criterion by which the shell of either sex can be distinguished. Out of nineteen specimens selected as males from my stock by persons professing to be able to distinguish the sexes, only one proved, on dissection, to be of that sex; and a small *U. pictorum* set aside as "undoubtedly female" turned out to be a male. I have always found males far less common than females.

[1] Cf. Baudon, *Rév. Mag. Zool.* 1852.

[2] von Jhering, *Zeit. wiss. Zool.* XXIX.; Schierholz, *ibid.* XXXI. 1878; and *Denkshr. Acad. Wiss. Wien*, LV. 1888–9; Goethe, *Zeit. wiss. Zool.* LII. 1891; Latter, *Proc. Zool. Soc.* 1891; Faussek, *Biol. Centralbl.* XV. 1895; and *5th Internat. Zoologic. Congress* (Berlin), 1901.

[3] Bronn, *Klas. u. Ord.* III. 1; Hazay, *Malacozool. Blät.* III. 1881.

The ovaries or testes, as the case may be, occupy the greater part of the visceral portion of the foot. Frequently the ovaries have an orange tinge while the testes are whitish: microscopic examination of their contents is, however, the only trustworthy means of determining the sex. The genital ducts pass forward, parallel with and ventral to the ureters, to the genital apertures which lie, right and left, immediately ventral to the nephridial opening.

At the spawning season, which takes place in May and June with *Unio*, but in June, July and August with *Anodonta*, the animals move towards the shallower, and therefore warmer, water. There is no actual union of the sexes: the spermatozoa of the males are simply discharged into the water and are carried by the ciliary currents into the inhalant siphon of the female. If a female be taken from the shell at this season the eggs may be seen through the transparent wall of the oviduct passing singly, but in a steady stream, to the genital aperture. Their motion is due partly to "labour contractions" of the intrinsic muscles of the foot and partly to the ciliated lining of the oviduct itself. One by one the eggs issue from the genital aperture, whence they are conveyed backwards by the abundant cilia which clothe the external surface of the nephridium. Along the middle line of this surface there is a belt of especially long cilia which appear to be devoted to the transit of the eggs; those dorsal and ventral to the belt work obliquely so as to keep the eggs in contact with it. It is probable that the free dorsal border of the inner lamella of the inner gill-plate is, under normal conditions, applied to the visceral mass in this region so as to enclose a temporary tube one of whose walls is formed by the above-mentioned belt of specialised cilia. In the course of about 50 seconds an egg is thus swept back to the slit between the protractor muscle of the shell and the point of fusion of the right and left inner gill lamellæ; here they meet the stream of ova from the other side of the body and so reach the exhalant current and the cloaca[1].

The process goes on for some 10 days or more in each individual and the number of eggs is immense. The estimates of various authorities range from 14,000 to 1,000,000; probably half a million may be taken as a fair average. On reaching the cloaca the eggs do not, as might be expected, pass out of the shell with the outflowing stream of water. On the contrary their direction is reversed and they pass forward into the cavities of the right and left outer gill-plates, which serve as brood pouches. The method by which this change of direction is accomplished is not quite clear. I have not been able to detect any reversal of the ordinary ciliary currents, nor can I accept the suggestion of von Baer, that the eggs accumulate in the cloaca and pass forward into the outer gill as the result of pressure due to their own accumulation. This theory necessitates the closure of the shell and consequent cessation of respiration during

[1] "Cloaca" is the term applied to a cavity common to the posterior portion of the digestive system and some other system or systems such as the excretory or genital or both of these. Here it is composed of the terminal parts of all systems which convey matter out of the body.

the process, and since the latter is prolonged for several days it seems probable that the animal would be suffocated.

I have, however, observed[1] on several occasions a violent and sudden reversion of the water-currents such as would certainly be fully capable of carrying the eggs forward and into the latticed recesses of the outer gills. This reversion is caused by the animal, firstly, closing all the ventral border of the shell by means of the free edges of the mantle assisted by the flexible, uncalcified rim of periostracum and leaving the siphons alone open, and secondly, relaxing the adductor muscles so as to allow the elastic ligament to make the valves gape apart. These actions cause the hydrostatic pressure within the shell to be less than that of the water without and consequently there ensues a rush of water into the shell through the open siphons. The whole procedure may be likened to a gulp and is achieved by precisely similar physical forces.

It may be enquired why the eggs do not also find their way into spaces of the inner gill. This is due partly to the fact that the space between the lamellæ of the outer gill-plate is greater than that between those of the inner and partly to the blocking of the latter by the following arrangement. Posterior to the foot the inner lamella of the outer gill-plate extends further towards the dorsal surface than the outer lamella of the inner gill-plate, and stretches over towards the middle line so as to nearly or completely meet its fellow

[1] Cf. Lloyd, *Ann. Mag. Nat. Hist.* 1870.

from the other side of the body and thus greatly diminish or even totally close the gap leading into the space within the inner gill-plates.

Each egg is contained in a transparent egg-shell from which, prior to fertilisation, there projects a chimney-pot-like tube, the micropyle. Fertilisation takes place within the gill-chamber by a spermatozoon swimming down the micropyle tube, which is thereupon closed and withdrawn; a wrinkled scar, resembling a "catherine-wheel," marking the position which it occupied. The egg then undergoes segmentation and at length the stage known as the "veliger" is reached. This very characteristic molluscan larva is provided with a belt of cilia, the velum. In the marine relatives of *Anodonta* the veliger escapes from the egg-shell and swims freely in the surface waters of the sea. In the present instance however the larva remains within the egg-shell, but nevertheless for a few hours its cilia become active, causing the animal to revolve slowly in the limited space at its disposal. It is remarkable that this evidence of a marine origin should still persist in a form that has for so long a period inhabited fresh water. Fossil *Anodonta* are found in the Palæozoic Rocks as far back as the Old Red Sandstone, the great antiquity of the genus is further attested by its present world-wide geographical distribution.

The veliger loses its cilia and by October the important stage known as the *Glochidium* is reached. The young animal now possesses a bivalve shell, each half of which is roughly triangular. The bases are united by a dorsal elastic ligament, and to each ventral apex is attached a sharp triangular tooth bearing numerous small secondary teeth upon its outer surface. The whole surface of the valves is pierced with numerous fine holes. An embryonal mantle lines the shell and bears on each side four steeple-shaped sense organs. A simple powerful adductor muscle is present, round which is coiled a sticky thread, the byssus, whose free end projects from a median papilla for a considerable distance.

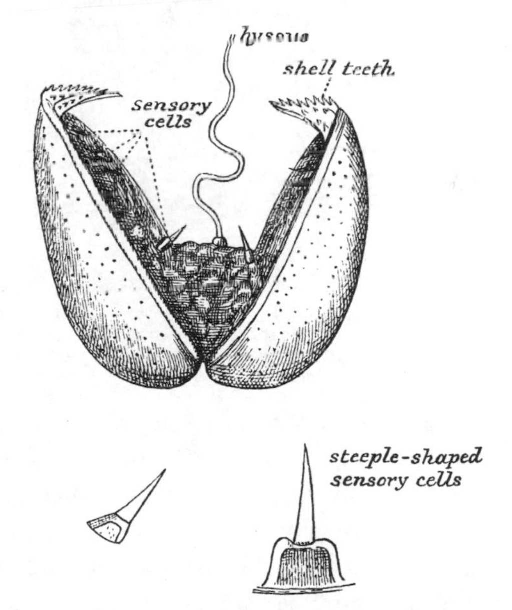

Fig. 34. Glochidium larva, as cast out by parent, viewed from behind.

Hitherto the developing larvæ have been both nourished and held securely within the brood-chamber, afforded by the outer gill-plate, by a nutritive mucus secreted by the epithelium of the junctions between the outer and inner lamellæ. Shortly after reaching the glochidium stage they emerge from the egg-shell and now maintain their attachment to the parent by means of their byssus threads.

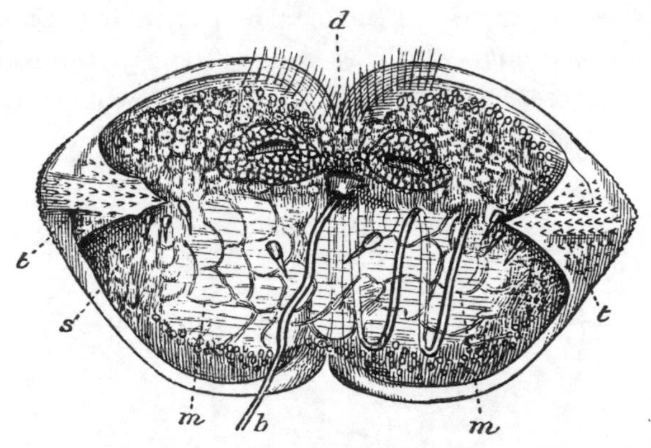

Fig. 35. Glochidium larva, ventral view.

b. Byssus thread, coiled round the adductor muscle, cut short. *d.* Future mouth. *m.* The adductor muscle showing through the mantle. *s.* Sensory cells. *t.* Main teeth and denticles on ventral margin of each valve.

It is not until the following February or March that the glochidia, measuring about 0·28 mm. dorso-ventrally and 0·35 mm. antero-posteriorly, are discharged from the brood-chamber into the surrounding water. The presence of a fish in the immediate neighbourhood, or any other gentle disturbance of the water, stimulates the parent mussel to discharge her young. I have observed that sticklebacks throng round a mussel when the glochidia are being discharged and appear much excited by the performance. They frequently snap at the entangled masses but spit them out again as though distasteful. Usually the young are sent forth in tangled cords through the exhalant siphon; occasionally singly and sometimes, according to some observers, through a small gap between the mantle lobes situated between the exhalant siphon and the elastic ligament.

So long as the slimy tangled cords of the glochidia remain attached to the parent, the latter is able to draw them back within the shell again by a kind of sucking action. The glochidia thus ejected are incapable of supporting an independent existence, neither are they able to swim. It is frequently stated that they do swim by clapping their valves together: this was long ago denied by Schierholz, and my own observations entirely confirm his denial. On leaving the shell of the parent they slowly sink to the bottom or are carried along by the stream, their byssi floating loosely in the water. If there are water weeds in the near neighbourhood the threads become spread like a spider's web among the leaves and branches. If however a fish comes near them they are thrown into a state of extreme excitement and clap their valves together with extraordinary vigour and rapidity: this may easily be seen by putting the tail or fin of a fish into a watch-glass containing freshly discharged glochidia. The effect of this valve-clapping is to force the byssus straight out from the shell. If now a byssus comes in contact with a fish it sticks to it and the whole tangled mass of glochidia is thereupon trailed after the fish.

Chance movements of the fins or tail are now almost certain to bring some of the glochidia into actual contact with the skin of the fish. Directly this happens the valves are snapped together, the apical teeth seize a portion of the skin and, folding within the opposed valves, drag it well down into the glochidial shell cavity and into contact with the mantle. The process is not unlike what takes place when a piece of soft material is caught between the teeth of two cog-wheels revolving in opposite directions. When once the glochidium has laid hold it never relaxes its grip. If a hard part, such as a spine of a stickleback, has been seized the glochidium soon dies and drops off; if however it is a soft portion, such as the gill filaments, or tail or lateral fins, to which it has become attached, then the tissues of the fish, irritated by the slight wound and the presence of the glochidium, become inflamed and rise up all round the seat of injury. In a short time a complete cyst is formed entirely covering the glochidium and rendering it impossible for the parasite, as it now is, to detach itself.

For the next three months the glochidium lives as a parasite upon its host, the fish. At first it obtains all its nourishment by intracellular digestion of the cells and juices of its host by means of its mantle. Subsequently the mantle undergoes a metamorphosis which results in its regeneration, the original cells being replaced by new ordinary ectodermic cells. Henceforth nourishment is obtained from the lymph streams of the host by means of amœboid cells situated in the gut of the young mussel. It is probable that the perforations of the glochidium shell are of importance in enabling diffusion to take place through it, especially during the earlier stages of parasitic life. Meanwhile the byssus and the adductor muscle of the glochidium disappear, and the two permanent adductors of the adult and the foot are developed. The mantle secretes a new bivalve and transparent shell resembling in miniature that of the adult and differing from that of the glochidium, which however is not as yet cast off but sits as a saddle upon the permanent shell beneath it. Eventually the cyst in which the parasite has been contained withers and the young mussel drops off and at length begins an independent life. The formation of the cyst and its subsequent degeneration are to be regarded as pathological phenomena occurring in the skin of the fish whereby it eventually succeeds in liberating itself from its parasite. Sticklebacks, loach, minnows, and probably many other species of fish are successfully affected by glochidia. Normal development also is said to take place when they are attached to axolotls or to newts. In my own experience however glochidia fail to secure a hold upon the skin of a newt. Tadpoles on the other hand succeed in freeing themselves from the glochidia before the latter have advanced far towards their adult structure.

When first set free the young mussel, as yet no larger than the original glochidium, is still covered by the glochidial shell, which persists outside the small permanent shell for three or four weeks and is the cause of the slight irregularity already mentioned (p. 163) in the lines of growth of the adult shell. The apical teeth of the glochidial shell project ventrally towards the middle line and as a consequence impinge upon the ventral border of the, at present, soft permanent shell at a point about half-way along its length. Hence at this point the permanent shell is prevented from growing so fast as elsewhere and therefore has its otherwise symmetrical curve sharply interrupted by an irregular notch pointing towards the

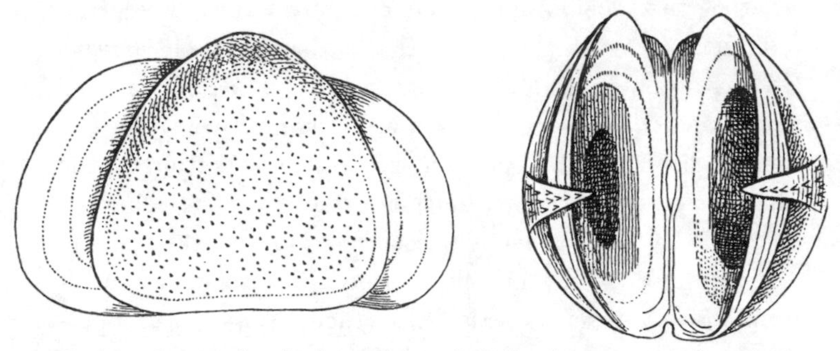

Fig. 36. A ventral and a lateral view of shells of young mussel at close of parasitic life, to show relation of permanent shell to glochidial shell. Magnified.

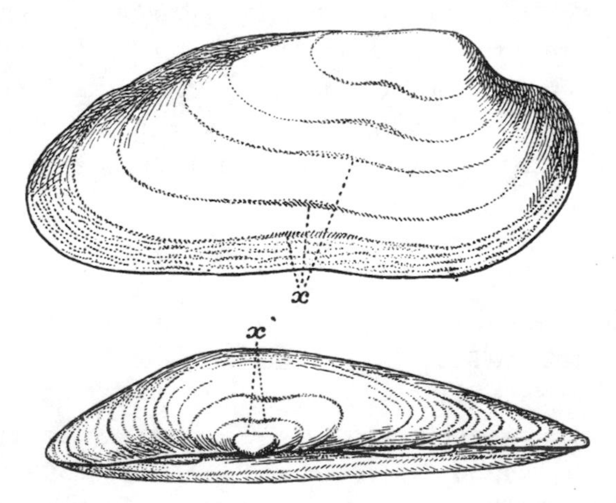

Fig. 37. A lateral and a dorsal view of a valve of shell of *Unio*, to show effect of glochidial shell-teeth upon lines of growth of mature shell. Reduced.

dorsal surface. This notch (x, Fig. 37) persists through life and causes the slight dorsal of the curves marking the lines of growth. In each successive line of growth the notch becomes of greater antero-posterior and less dorso-ventral extent, thus tending to become less evident and to disappear. It can therefore be most easily seen near the umbones of those shells which have escaped corrosion.

The foot of the young mussel is blunt and far flatter than that of the adult. When the animal is advancing, the foot at first protrudes slowly until it stretches a con-

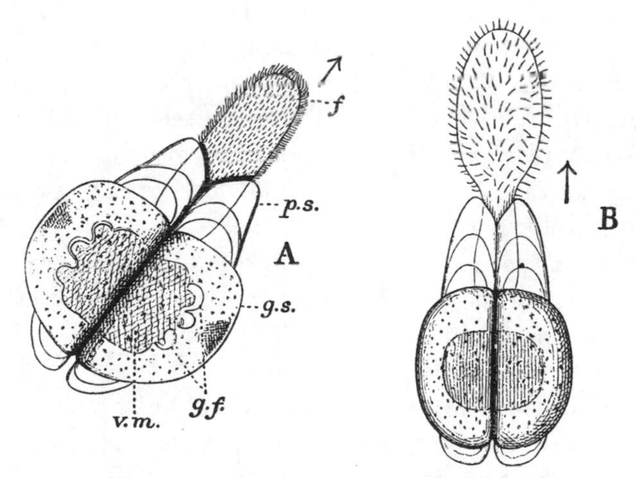

Fig. 38. Young mussel about seven weeks after close of parasitic life and 129 days after first attachment to stickleback, with glochidial shell as a saddle over the delicate permanent shell. In A the animal is slowly pushing the ciliated foot forward in the direction indicated by the arrow. In B the foot is fixed, the cilia motionless and the shell is being drawn forward, its valves are drawn together by the adductors so as to pinch the hinder part of the foot.

f. Foot. *g.f.* Three first gill filaments seen through shells. *g.s.* Glochidial shell. *p.s.* Permanent shell. *v.m.* Visceral mass. Magnified.

siderable distance in front of the shell, the cilia on its surface moving all the while with great rapidity; the shell is then drawn forward over the foot by a rapid muscular contraction. Even when crawling over a smooth microscope-slide the foot does not slip back but remains stationary when this action takes place. As soon as the contraction commences the cilia suddenly cease moving and stand out from the surface of the foot, like the bristles of a brush, absolutely motionless and rigid: in this condition they remain until the foot again begins to glide forward. It is possible that the cilia which are in contact with the glass, or other surface, may actually adhere to it and thus fix the foot. The appearances suggest very forcibly that, at the moment of drawing the shell forward, the pressure within the tissues of the foot becomes so great that the cilia cannot assume any other position than one perpendicular to the surface.

Nothing is known of the growth and adolescence of the young mussels; their minute size and transparency render them exceedingly difficult to capture even in the limited area of an aquarium.

Powers of Endurance. It is interesting to find that mussels are able to withstand severe cold, even to the extent of freezing and this even after removal from the shell. I have seen an *Anodonta* frozen solid on two successive nights as it lay in the dissecting dish and yet return to life, as manifested by pulsations of the heart and activity of the cilia, when slowly thawed. This is true not only of the adult animal but also of the glochidia within the gills. This power cannot fail to be of great value in preserving the species in many shallow ponds which are frequently frozen to the bottom in severe weather. Exposure to the air is tolerated for several hours provided the animals are kept cool and not exposed to a hot sun. Warming the water to a moderate degree will often cause protrusion of the foot and promote general activity; excessive warmth however has the reverse effect and is speedily fatal.

Enemies, Parasites, etc. *Anodonta* is a favourite article of food among coot, several species of wild duck and the heron; the latter at times carries them away to the heronry and there breaks open the shell against the branches. Occasionally water-voles appear to eat them, for empty shells are sometimes found in quantities near their burrows. Otters too are recorded as preying upon them, nor is the "bargee" above making a meal at their expense.

The corrosion which is so frequently to be noticed upon the outer surfaces of the shells, especially in the region of the umbones, is the result of the attacks of a unicellular alga *Micrococcus conchivorus* with which are often associated species of *Batrachospermum* and others.

It is hardly possible to dissect an *Anodonta* without finding on the surface of the mantle or foot or gills one or more specimens of a small black and white acarid, *Atax bonzi*[1]. This and another species, *A. crassipes*, that is sometimes also found associated with the mussel, is a commensal rather than a true parasite. *A. bonzi* passes its entire life in connexion with *Unionidæ*, leaving one

[1] Claparède, *Zeit. wiss. Zool.* XVIII. 1868.

individual only to seek another: occasionally it may be seen running about upon the outside of the shell. *A. crassipes* on the other hand quits its host after reaching the second larval stage and lives free when adult. The females of both species deposit their eggs in the gill tissue of the mussel, and it is in the gills that the larval metamorphoses take place, but in the intervals between the ecdyses the larvæ quit the gill chambers and wander more freely about the body.

The eggs of other parasites have been recorded from other parts of the body, viz., of *Hydrocharis* in Keber's organ and of *Aspidogaster* in the pericardium. The adult *Aspidogaster* is also found in the same cavity. The most interesting of the parasitic organisms are however the various species of Trematode Flatworms which pass their immature stages within the body of the mussel, but only become mature on reaching a second host, probably an aquatic bird. The sporocysts and cercariæ are found in the spaces between the gut and the genital glands and also in the mantle and pericardium, but the complete life-history has not yet been worked out in any one instance. The suggestion just put forward is based upon the analogy of what is now known to be the case in the pearl-forming sea-mussel, *Mytilus edulis.*

Pearls. Pearls may be formed in almost any of the tissues of the mussel, either singly or in groups; they may be loose in the tissues or may become attached to the inner surface of the shell. In all cases they consist of alternating concentric layers of calcareous matter and conchiolin, being in fact identical in structure with the innermost or nacreous layer of the shell. The determining cause of the formation of a pearl is the presence of some foreign body which serves as an irritant and forms, as it were, a nucleus round which the nacreous deposit is laid down. The pearly matter is produced at the expense of the shell, and hence pearl-bearing specimens are usually to some extent deformed or misshapen. Dr Lyster Jameson[1] has recently shown that in the case of the marine mussel *Mytilus edulis* pearls are formed in a closed sac of the shell-secreting epithelium of the mantle, and that those formed in the extreme margin of the mantle are composed mainly of periostracum, while those which occur elsewhere are composed of nacreous material. The epithelial sac is first formed round a living trematode larva, *Distomum (Leucithodendrium) somateriæ* Lev., which enters upon a resting stage in the tissues of *Mytilus*. The trematode may become imprisoned within the pearl and die or may effect its escape, but, the cyst having been formed, the pearl is in either case completed. The first host of this trematode is the cockle, *Cardium edule*, or the "tapestry shell," *Tapes decussatus*; the final hosts to which it gains access by the *Mytilus* are almost certainly the eider-duck, *Somateria mollissima*, and the black scoter, *Œdemia nigra*, both of which birds devour quantities of *Mytilus*. It is highly probable that the pearls found in the *Unionidæ*, notably in the Scotch river-mussel *Unio margaritifer*, will, if investigated, prove to be of like origin and causation.

Pearl fisheries are still existent in various Irish and

[1] *Proc. Zool. Soc.* vol. I. London, 1902.

Scotch rivers, and occasionally pearls of high value are found in *Unio margaritifer*, which is there abundant. In past times the Tay fishery was very productive and is stated to have yielded pearls to the value of £10,000 in the three years 1761—1764.

The origin of pearls in the pearl oyster *Margaritifera vulgaris* (*Avicula fucata*) has been traced to a similar source. Kelaart first, in 1859, connected the formation of pearls with parasitic worms. Recently Prof. Herdman[1] has made an investigation of the Ceylon pearl fisheries and finds that nematodes, trematodes and cestodes are all concerned, while rarely a grain of sand forms the "nucleus" in pearl formation. The most common cause of pearls he finds to be a larval cestode of a *Tetrarhynchus* form. In its earliest stages the larva is said to be free swimming: from the oyster it probably passes into the file-fish (*Balistes* sp.), or into some species of *Trygon* which prey upon the oysters.

Before closing this chapter it will be well to review a few noteworthy features in the life-history of the mussel. When compared with that of marine Lamellibranchs the life-cycle is remarkable on two accounts; firstly, the prolonged period of attachment to the parent, and the suppression of the free-swimming veliger stage; secondly, the post-embryonic life as a parasite upon the skin of a fish. In the majority of marine molluscs dispersal is effected during the veliger stage, the powers of locomotion of the adults being in most cases feeble or altogether absent. The minute transparent veliger swims in the

[1] *Nature*, vol. LXVII. April 30, 1903.

surface waters and is carried hither and thither by the waves and tidal currents. This mode of dispersal is however impossible for an animal inhabiting running fresh-water, for the veligers would inevitably be carried along by the stream and in time swept out into the sea. Here then we see a cause which has led to the retention of the young within the brood-chamber of the parent. Other fresh-water animals exhibit the same peculiarity; many species not only of molluscs, but also of crustaceans, worms and cœlenterates, either by maintaining the young attached to the parent or by stocking the egg with abundant food material, defer the acquisition of a free life by the embryo to a far later period than is the case with their marine relatives. Exceptions exist, but it would be going beyond the scope of this book to discuss them. There is little doubt that the prevalence of parental care for the young among the fresh-water fauna is the result of running streams making their way to the salt waters of the sea. An interesting exception is afforded by *Dreissensia* (*Dreissena*) *polymorpha*. This species inhabits both brackish estuarine waters and fresh. It has a free-swimming veliger larva which certainly could not make headway against a stream. Nevertheless the species has spread up the Thames and into the Oxford and Birmingham Canal, while on the Continent it has travelled up the Rhine, along the Main and Danube Canal and into the Danube. These extensions of its distribution are of recent date and are in fact due to human intercourse. The adult *D. polymorpha* fixes itself by its byssus to various, relatively firm, objects, such as logs and the bottoms of boats, and in this way has been conveyed against the stream along the waterways of commercial traffic.

Anodonta and the other *Unionidæ* having been fresh-water animals for long ages, have availed themselves of the superior activity and power of dispersal possessed by fish. The glochidia when attached to their host may be carried many miles from the place of their birth. It is even conceivable that they may have been transported from one stream to another by such fish as make periodic journeys to the sea. It would be extremely interesting to try how far the glochidium during its parasitic life is tolerant of salt water. The experiment could be conducted with some species of stickleback (*Gasterosteus*) which live indifferently in fresh or in brackish water.

BRITISH FRESH-WATER LAMELLIBRANCHS.

Class Lamellibranchia (Pelecypoda).

Order Eulamellibranchia.

Family Unionidæ. Shell equivalve, oblong, inequilateral, pearly within; large external ligament; anterior hinge-teeth thick and striated, posterior teeth often absent; all hinge-teeth occasionally rudimentary; adductor muscle-impressions deep.

genus *Unio*, hinge-teeth strongly marked, lunule (depressed area in front of umbo) distinct, shell thick, solid.

U. margaritifer (L.), $5'' \times 2\frac{2}{5}''$, oblong, lower margin straight.

U. pictorum (L.), $3'' \times 1\frac{1}{3}''$, oblong, umbones with small tubercles not confluent, anterior teeth compressed and crenulated.

U. tumidus (Retz), 3″ × 1½″, oval, umbones with angular wrinkles, confluent and concentric, lower margin curved.

genus *Anodonta*, hinge-teeth rudimentary or absent, lunule indistinct, shell thin.

A. cygnea (L.), 5⅓″ × 2¾″, oblong, umbo at one-fourth from anterior end; muscular impressions indistinct. *A. anatina* is but a variety of *A. cygnea*[1].

Family SPHÆRIIDÆ. Shell suborbicular and thin; hinge with cardinal and lateral teeth; ligament external and placed on posterior side of hinge; foot large and tongue-shaped, no byssus, one siphon, or two siphons more or less united.

genus *Sphærium* Scop., shell nearly equilateral, umbo near middle of dorsal margin; two siphons.

S. corneum (L.), ⅖″ × 3/10″, yellow horn-colour with faint concentric bands. In June, 1901, masses of this species blocked the inflow of the bathing tank at Christ's College, Cambridge.

S. lacustre (Müll.), ⅖″ × 3/10″, yellowish-white or ashy-grey, thin, compressed, sides truncated and sloping from dorsal margin.

S. pallidum Gray, ⅗″ × ⅖″, drab or yellowish, anterior side rounded, posterior side truncated.

S. rivicola (Leach), 9/10″ × 7/10″, horn-coloured with dark concentric bands and deep ridges, ligament very conspicuous.

genus *Pisidium* (Pfeiffer), shell inequilateral, one siphon.

P. amnicum (Müll.), ⅖″ × ⅓″, shell triangular, inequilateral, with the anterior side the longer.

P. subtunicatum Malm, ⅙″ × 1/7″, shell triangular, striated concentrically, posterior side much produced and rounded; var. *henslowana* has plate-like appendage to umbo.

P. nitidum Jenyns, 1/7″ × 1/7″, shell round, anterior side truncate and rounded, posterior side produced and sloping abruptly.

P. pusillum (Gmelin), ⅕″ × ⅙″, shell oval, both sides rounded and compressed.

[1] Woodward, "List of British Non-Marine Mollusca," *Journal of Conchology*, October 1903, p. 352.

ORDER FILIBRANCHIA.

Family MYTILIDÆ. Byssus well developed: anterior adductor muscle small.

genus *Dreissensia* (van Ben), shell equivalve, inequilateral, swollen, umbones anterior terminal; below umbo a triangular shelf for support of anterior adductor muscle; ligament internal; hinge toothless or with minute cardinals; anterior adductor muscle impression small, posterior large. Siphons prominent.

Dreissensia (*Dreissena*) *polymorpha* (Pall.), 1¼″, shell sea-mussel-shaped and keeled in the middle of both valves.

CHAPTER VII.

SNAILS AND SLUGS.

THE fauna of the British Isles is fairly rich in the terrestrial and fresh-water Gastropod Molluscs popularly known as Snails and Slugs. The shell of the common garden snail, *Helix aspersa*, or of the larger *Helix pomatia* may be taken as typical. It is spirally coiled, and marked with numerous fine lines of growth running parallel with the margin of the matter of the shell and indicating successive positions of the free edge during the growth. The surface is decorated with several bands of colour, whose arrangement and pattern are liable to great variations. A spiral "suture" runs from the tip or "apex," round and round the shell to the upper edge of the mouth: this suture marks the line along which the growing margin of the shell has become fused with the already existing whorls. It will be seen that the direction of growth has always been towards the animal's right side: the shell exhibits a "dextral" or right-handed spiral. This will be more easily grasped if the reader imagine himself to be on the tip of the shell, in its natural

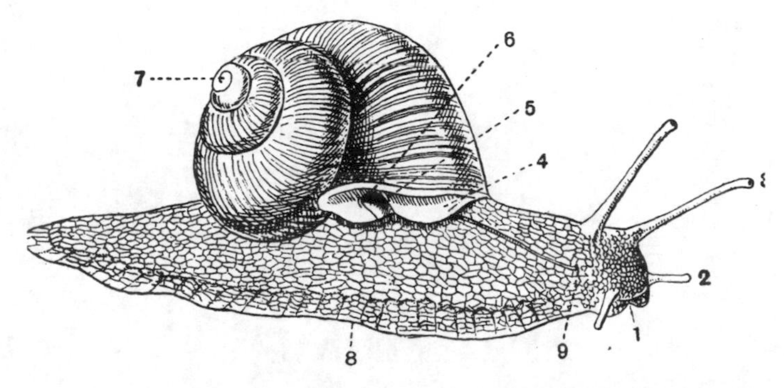

Fig. 39. *Helix pomatia.* Side view of shell and animal expanded. From Hatschek and Cori.

1. Mouth. 2. Anterior tentacles. 3. Eye tentacles. 4. Edge of mantle. 5. Respiratory pore. 6. Anus. 7. Apex of shell. 8. Foot. 9. Reproductive aperture.

position, and about to walk down along the suture; he will then keep constantly turning towards his right hand. The majority of gastropod shells are dextral; occasionally sinistral examples are found of species normally dextral, *e.g.* the common whelk, while in some few the sinistral condition is the rule.

On the under surface of the shell is a small opening, the "umbilicus," leading into a hollow pillar, the "columella," which is formed by the contact of successive whorls of the shell. The umbilicus is completely hidden by a reflected portion of the shell-lip in *H. aspersa*, but in *H. pomatia* it is partly exposed.

The "apex" is the oldest part of the shell and was at the first sufficient to contain the entire young snail. As the animal grew a larger shell became necessary, and accordingly additions were made to the free lip of the small original shell, *i.e.* to what is the apex of the fully grown shell. This process is repeated as often as may be necessary and gives rise to the lines of growth already mentioned. By far the greater part of the shell is formed by glands situated near the thickened edge of the "mantle," known as the "collar." The details of the mode of formation have been worked out by Longe and Mer[1], and independently by Moynier de Villepoix[2]. In growing snails there may be seen just behind the collar a white band, the pallial gland. In front of this is a groove into which glandular structures open; it is here that the outer layer or "epidermis" (periostracum) of the shell is produced. This layer is horny and uncalcified and is best seen round the shell-lip in young snails during periods of active growth, where it is almost free from the colours which it subsequently acquires, and is quite flexible, being as yet destitute of the subjacent harder calcareous constituents of the older shell. The white band secretes the dense calcareous matter of the middle layer of the shell, and the surface of the mantle posterior to the band completes the process by adding the innermost or "nacreous" layer. When a snail is fully grown the epidermic groove and the white band gradually disappear, but the general surface of the mantle retains the power of laying down shell material. Normally, and in the absence of accident the additions thus made to the thickness of the shell are comparatively slight. The shell increases in thickness from the apex towards the lip, shewing that the chief additions are those

[1] *C. R. Acad. Sci.* Paris, xc. 1880.

[2] *Ibid.* CXIII. and *Bull. Soc. Zool. Fr.* XVII.

made by the glandular organs just described. If however a portion of shell is cut out from the midst of one of the whorls, within two hours a thin organic membrane containing crystals of calcium carbonate is secreted and rapidly closes the hole. By subsequent internal additions the new piece is thickened and made as strong as the surrounding uninjured parts but it is devoid of periostracum. On the other hand if the injury is inflicted upon the shell-lip of a growing snail both the outer non-calcareous "periostracum" and the internal calcareous portions are alike repaired. The calcareous matter is in the form of that modification of calcium carbonate known as aragonite.

The normal rate of shell-growth probably varies considerably according to climatic conditions, the nature and amount of food-supply, and so on. According to Lowe[1] the shell increases but little for a considerable time, and never becomes mature before at least one period of dormancy during which no growth occurs. Many species are said by him to burrow in the ground and conceal themselves during periods of shell growth. In this opinion he is at variance with Collinge[2]. Be that as it may, there is no doubt that when growth does occur it is very rapid. *H. variabilis*, *H. pisana*, and *H. aspersa* have been known to add more than a centimetre of shell in four or five days. Nearly all snails' shells exhibit in addition to the lines of growth a few well-marked lines some distance apart and often separating areas of different

[1] *Phil. Trans.* 1854.

[2] *Naturalist*, 1891 ; *Conchologist*, II. 1892.

depths of colour. These are usually regarded as distinguishing one year's growth from that of another. This explanation is very probably correct, but I am not aware of any exact observations on the matter. The length of life of *H. aspersa* is said to be about five years, and that of *H. pomatia* from six to eight years. It is however highly probable that growth ceases some time before natural death ensues; hence shells could hardly be expected to exhibit a number of "annual lines" exactly corresponding to the years of life of an old snail. I have never found more than three "annual lines" on shells of *H. aspersa.*

The thickness of the shell is dependent upon the nature of the soil in any given locality. Thus Taylor states that shells of *H. aspersa* collected in Guernsey where calcareous strata are absent are very thin, whereas those found on the limestone strata at Swanage and Tenby are unusually thick.

Semper's[1] researches on the phenomena of growth have shown that in the freshwater snail *Limnæa stagnalis* the size attained by the shell is influenced both by the temperature and by the volume of the water in which the specimens are reared. The optimum temperature for this species is between 68° F. and 77° F., while growth ceases alike if the temperature exceed 90° F. or fall below 53° F. The volume of water required per individual is about 5000 cubic centimetres. Specimens reared in 100 c.c. of water attained a length of only 6 mm. in 65 days, others, members of the same family, reared in 250 c.c. of water became 9 mm. long, while yet others in 2000 c.c. of water

[1] *"Animal Life," Internat. Sci. Ser.* vol. XXXI.

reached a length of 18 mm. in the same period. It is perhaps hardly necessary to point out that the amount of calcareous matter dissolved in 2000 c.c. of water is greater than that in 100 c.c. This is probably an important factor in determining growth. It would be interesting to vary Semper's experiments by dissolving (either at once or gradually during the course of experiment) in 100 c.c. of water as much calcareous matter as is actually contained in 2000 c.c. of the same water. This would, *ceteris paribus*, enable us to determine the influence of volume alone upon the size of the shells.

The *Helicidæ*, *Limnæidæ* and other air-breathing or (pulmonate) gastropods have no lid (operculum) by which the aperture of the shell can be closed when the animal has retreated within. Such lids are well known in the common periwinkles and whelks of our shore pools and may also be seen upon the fresh-water snails *Vivipara* (*Paludina*), *Bithynia* and others; they are horny, sometimes calcified structures developed upon the "foot" of the animal; they are marked with concentric lines caused by additions to the margin as need arises for a larger operculum to fit the increasing shell-mouth. The land snails however can produce somewhat analogous structures at certain times.

When a snail hibernates it retires into some sheltered spot or even excavates a tunnel in the earth or rock (*e.g.* *H. aspersa* in limestone rocks) and closes the aperture of the shell by a disc, the "epiphragm[1]" or "hibernaculum,"

[1] Keferstein, *Bronn's Klass. u. Ord.* III. 2. Binney, *Terr. Moll. of U. S.* II. 1851. Barfurth, *Arch. f. Mikrosk. Anat.* XXII. 1883. Allman, *Jour. Linn. Soc. Zool.* XXV.

composed of organic and calcareous matter. When the animal is about to form the epiphragm it retires within the shell and exudes from the collar of the mantle a quantity of mucus charged with calcareous matter. With this the mouth of the shell is filled. A small quantity of air is then discharged from the respiratory aperture and separates the film of calcified mucus from the body of the animal. The pressure of the discharged air causes the still flexible epiphragm to bulge outwards. Almost at the same moment the animal retreats further within the shell and the pressure of the outside air forces the epiphragm back, making it flat or even concave[1]. The whole structure then speedily hardens, but remains at all times porous, so that air can diffuse through. It is however insoluble in and but slightly pervious to water. At times several epiphragms are found one within the other as the snail retreats farther and farther into its shell: in such case each succeeding epiphragm is more delicate and less calcareous than its predecessor.

During periods of drought a "summer epiphragm" is frequently formed. This is a far more delicate structure than that just described, being usually transparent and but feebly calcified: it often exhibits a white calcareous spot opposite the respiratory aperture, and this spot is usually perforated. Probably a small aperture for respiration exists in all epiphragms, both "winter" and "summer." Some species of fresh-water snails, notably of the genus *Planorbis*, close the shell-mouth by a firm white epiphragm when the streams and pools in which

[1] Binney, *Bull. Mus. C. Z. Harv.* IV.

they live dry up; and this in addition to burying themselves in the mud. A similar but permanent and moveable contrivance for closing the aperture of the shell is found in the pulmonate gastropods *Clausilia.* This structure is known as the "clausilium." It consists of a white plate attached by an elastic spirally-twisted stalk to the columella. When the animal comes out of the shell it presses the clausilium back against the inside of the shell, leaving the exit free. On the retreat of the animal the elastic stalk pulls the clausilium into position across the aperture. The whole may be compared to a door shut with a spring. There can be no doubt that in all cases the purpose of the epiphragm is not only, nor perhaps chiefly, to form a protection against climatic conditions, but also to prevent the entry of carnivorous foes such as beetles, crustaceans, and parasitic worms. The hibernation of slugs generally takes place below the surface of the earth within a slimy cocoon.

The body of *Helix* is attached to the shell by means of the columellar muscle which runs along the inner side of the spiral of the shell and is fastened by tendons to the columella in the upper part of the first turn. From the tendinous origins strong bundles of muscle fibres pass down into the foot: it is by the contraction of these that the animal is retracted within the shell. Slugs, as is well known, have not a similar protective shell and, indeed, are popularly supposed to have none. Nevertheless the shell-sac of *Arion* (*A. aier,* the large black slug) contains numerous crystals of calcium carbonate, and in *Limax* (*L. flavus,* the yellow cellar slug, and *L. maximus,* a large pale grey slug spotted with black) a small shell is present. The carnivorous slugs of the genus *Testacella* carry on the hinder part of the body a cap-like shell with evident traces of spiral growth. It is thus clear that slugs are descended from ancestors that possessed well-developed shells within which the animal could be retracted.

The locomotion of these Molluscs is proverbially slow. Slugs are less tardy than snails and have been timed to travel at the rate of a mile in about eight days, whereas a snail (*H. aspersa*) would take nearly fifteen days to cover the same distance[1]. The movement is a steady, gliding one and is brought about partly by muscular and partly by ciliary action. From the pedal gland, whose aperture is just below the mouth and above the anterior end of the foot, mucus is discharged on to the surface over which the animal is travelling: the discharge is effected by the cilia, which line the cavity of the gland, assisted by muscular compression. On the smooth bed of mucus thus laid down the animal glides forward, advancing by the aid of the cilia which cover the sole of the foot (and the sides in some slugs, *e.g. Arion*) and by a series of successive waves of muscular contraction and expansion flowing over the sole from behind forwards. These waves may vary in number from some 30 to 50 per minute. The slimy trail left behind by the mollusc is marked with transverse wavy ridges and furrows about a millimetre apart as the result of the undulatory movements of the foot. Tracks of a more permanent kind are often made by snails or slugs gnawing the surface over which they are moving; such

[1] Jeffreys (Thomas), *Brit. Conch.* i.

tracks have been found on the glass roofs of greenhouses from which snails have removed the whitening, and may be seen on the bark of trees over which they have crawled. These last are sinuous bands from which the microscopic Algæ, *Pleurococcus*[1], (the "green" of the bark) have been removed by the browsing snail.

The slime trails both of slugs and snails show clearly that these animals possess a good sense of direction and locality. Each individual as a rule returns after its nocturnal forays to its own selected retreat, whence it issues again the following evening.

Aquatic snails such as *Limnæa* glide along by similar means in search of food over the surface of water plants or in an inverted position just below the surface of the water with the sole uppermost[2]. In this position advantage is taken of the physical properties of the "surface-film" to procure a relatively firm basis upon which the mucus can be deposited.

Many fluviatile molluscs, such as *Limnæa*[3], *Physa*, *Planorbis*, *Cyclas*, and others, possess a more rapid method of locomotion in a vertical plane. This consists in the power of forming threads of mucus of considerable length either in an upward or downward direction. These filaments may be attached to objects below the surface of the water or to the "surface-film" itself. In the latter case the upper end of the filament is expanded and slightly concave, and acts as a float supporting the thread below it. The

[1] *Feuille de Jeun. Natural.* (3) Ann. 28.

[2] Gräfin M. von Linden, *Biol. Centralbl.* xi. 1891.

[3] Warington, *Ann. Mag. Nat. Hist.* 1852.

snails use these cords as a convenient and direct route to the surface for respiratory purposes or for a quick return to the depths when a fresh supply of air has been taken in. A *Limnœa* in climbing up its cord often folds the foot longitudinally so as to approximate the right and left edges and enclose the cord in a temporary tube. This manœuvre brings the whole ventral surface of the foot in contact with the cord. At the same time muscular movements of the body are visible which strongly suggest that the animal is pulling itself upwards. It is nevertheless most probable that in these relatively rapid ascents and descents hydrostatic principles are chiefly concerned. I have seen a *Limnœa* in an aquarium ascend from a depth of eight inches with a bubble of air projecting from the respiratory aperture. On arrival at the surface the animal rolled over so as to bring the aperture to the surface, and the bubble burst with an audible explosion. The lips of the aperture were then projected funnel-wise above the surface for some seconds while fresh air poured in; this was repeated several times, the aperture being closed after intake of air and lowered just beneath the surface. Having obtained enough fresh air the *Limnœa* descended again, alighting, thanks to the guide rope, upon the exact spot whence it had started. The rise and fall chiefly depend, as I think, upon the amount of water displaced by the snail and its contained air: the animal is a living Cartesian diver.

A similar power of forming threads[1] is possessed by many slugs, especially when young, and enables them to

[1] Zykoff, *Zool. Anz.* 1889; Martins, *ibid.* 1878; Eimer, *ibid.* 1878; Tyd. *Quart. Journ. of Conch.* 1878.

descend quickly through the air from trees and shrubs to the ground or if necessary to perform the return journey. During descent the slug travels head downwards, paying out fresh mucus behind it, and gripping the cord with the sides of the foot. The upward journey is of course a more laborious task and is performed by bringing head and tail together and transferring the point of attachment to the former; the animal now travels head uppermost and the gathered in cord is accumulated near the tail. In *Limax maximus*, and some other species, pairing habitually takes place in mid air; the two individuals suspend themselves head downward from some suitable object by a double cord of mucus which supports their intertwined bodies.

Food and Digestion. Plants form the chief food-supply of terrestrial and fresh-water molluscs: indeed the mechanical and chemical protective devices of wild plants seem to be specially directed against the attacks of snails and slugs. Tender succulent shoots and saccharine portions of the plants are preferred and unerringly selected. Slugs however are very general feeders; some species of *Limacidæ* are omnivorous (I have seen a *Limax* devouring bird's droppings), others confine themselves to fungi; all are occasionally carnivorous, and some are predatory and even cannibalistic; the species of *Testacella* habitually devour earth-worms and other small animals. The food, whatever its nature, is attacked by means of the radula upon the floor and the hard crescentic chitinous jaw upon the roof of the mouth. The radula[1], or lingual ribbon, is a

[1] Geddes, *Trans. Zool. Soc. Lond.* x. Rössler, *Zeit. wiss. Zool.* XLI. 1885. Loisel, *Journ. de l'Anat. et Physiol.* XXVIII. Rücker, *Bericht*, 22 *d. Oberh. Gesellsch. f. Natur. u. Heilkd.* Giessen.

horny sheet covered with an immense number (about 15,000) of small, backwardly directed teeth; it is carried upon the subradular membrane, which in turn is borne upon cartilaginous pads capable of rotating through a small angle. The whole is worked by a system of protractor and retractor muscles. The action may be most conveniently observed in aquatic gastropods crawling upon the glass of the aquarium and browsing on the microscopic algæ that adhere to the glass. At each bite the upper lip is retracted and the jaw brought close to the glass, the radula is then pushed forward and makes an upward rasping stroke: by this movement the algæ are torn off and rasped against the jaw. The radula of *Testacella* is capable of rapid protrusion, the prey being impaled upon the teeth and drawn into the mouth, which in this case is not furnished with a jaw. The anterior portions of the radula are continually being worn away by friction but, as continually, are replaced by fresh growths from behind. The growth takes place in the radula sac, which is a ventral offset from the hinder part of the mouth cavity, and the whole horny sheet slides slowly forward over the floor of the mouth, keeping up an unending supply of fresh teeth throughout life.

The food in the mouth is exposed to the action of the secretion of the salivary glands. This fluid not only moistens the food to assist its passage along the alimentary canal but also converts the starches of the food into sugar[1]. The lining of the canal is more or less provided with cilia, whose action, combined with the peristaltic contractions of

[1] Bonardi, *Boll. Sc. Pavia*, 1883.

the external circular and internal longitudinal muscular coats urges the food onwards. Further chemical changes are effected in the food by the secretion of the digestive

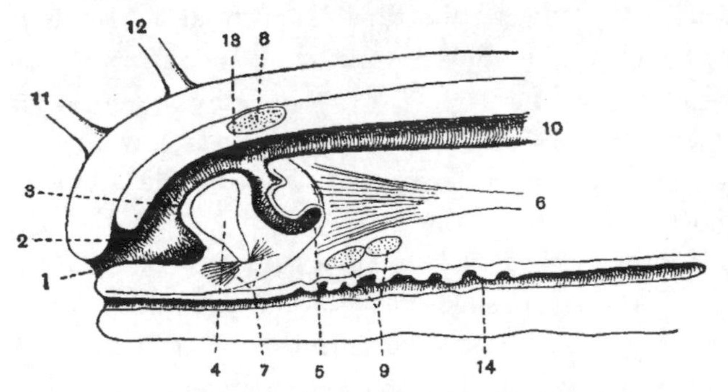

Fig. 40. Inner view of right half of head of *Helix*, to show the arrangement of the Radula × 2.

1. Mouth. 2. Horny jaw. 3. Radula. 4. Cartilaginous piece supporting radula. 5. Radula sac from which radula grows. 6. Muscle which retracts the buccal mass. 7. Intrinsic muscles which rotate the radula. 8. Cerebral ganglion. 9. Pedal and visceral ganglia. 10 Œsophagus. 11. Anterior tentacle. 12. Eye tentacle. 13. Orifice of duct of salivary gland. 14. Mucous gland which runs along foot and opens just under the mouth.

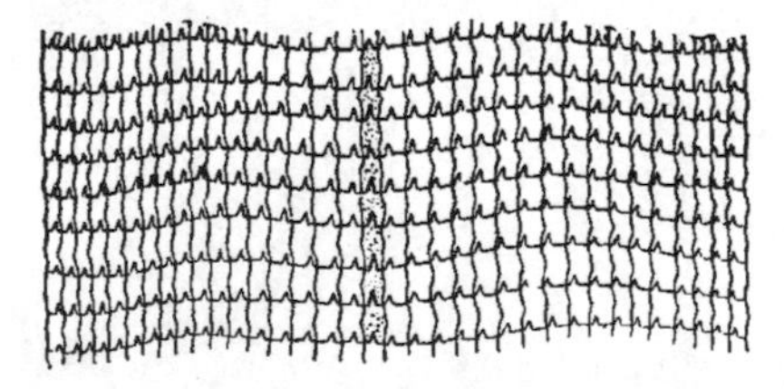

Fig. 41. A small portion of the radula of *Helix*. The central row of teeth has been shaded. Magnified. (After Howes.)

gland or "liver." Our knowledge of the action of this gland is still incomplete. It is known that its secretion is acid and contains both diastatic and peptic ferments and is capable of emulsifying fats. There are also found in it hæmatin and a product of the decompositions of chlorophyllaceous food termed "entero-chlorophyll." In the gland itself Barfurth[1] distinguished ferment-, lime- and liver-cells. Mac Munn[2] has shown that granules may be present both in ferment- and in liver-cells and regards the granules as being entero-chlorophyll. In this he is supported by Miss Newbegin[3], who however holds that the excretion of the granules is the work of the epithelium of the intestine. By Biedermann and Moritz[4] the granular (liver) cells are regarded as absorptive, this function being but sparingly performed by the intestinal epithelium. In view of results obtained by other investigators of the functions of the digestive glands of other invertebrata, there is much probability in the accuracy of this view. Glycogen is found in the connective tissue cells of the gland in *Helix* and in the granular (liver) cells themselves in *Limax*. The amount of lime present in the digestive gland varies at different seasons. Barfurth found that in the digestive gland of *H. pomatia* the percentage of inorganic ash averaged 20·24 in May, 25·72 in September, 10·26 in winter after the formation of the epiphragm, and 16·99 after the breaking and repair of the shell. It is clear then that in addition to its other functions the gland serves as a storage organ for calcareous matter. The fæces are extruded at the anus by the action of a longitudinal band of muscle upon the outer side of the rectum:

[1] *Op. cit.*

[2] *Phil. Trans.* CXCIII. B.

[3] *Q. J. M. S.* XLI.; and *Zool. Anz.* XXII.

[4] *Pflüger's Arch. f. ges. Physiol.* LXXV.

when this muscle contracts the rectum is thereby shortened and its contents driven out, usually with a decided spiral movement.

Circulation and Respiration. The blood of *Helix* performs both respiratory and nutritive functions[1]; accordingly we find the circulatory system well developed. The heart beats at the rate of about 58 pulsations per minute in an adult *Helix hortensis,* but more rapidly in young individuals. The rate is however variable in accordance with temperature, muscular exertion, and similar influences. The auricle of the heart receives aërated blood from the pulmonary vein and drives it on into the muscular pear-shaped ventricle. The passage from auricle to ventricle is guarded by a valve which prevents the blood from flowing back to the auricle. From the ventricle the blood passes into the main aorta, which after a short course bifurcates into the anterior and posterior aortæ: the former of these distributes blood to the salivary glands, anterior part of the foot, buccal mass and head; the latter to the other digestive and to the reproductive organs. The branches of these main arteries are comparatively few and debouch into irregular spaces among the tissues. These spaces eventually unite either with the large visceral sinus which runs from the apex of the visceral coil, or with one of the two sinuses lying at the sides of the pedal gland. From these large sinuses the blood, now charged with carbon dioxide gas and to a great

[1] Milne Edwards, *Mém. Acad. Sci. Paris*, XX.; Girod, *2nd Internat. Congr. Zool. Moscow*, Part 2, 1892; Natepa, *Sitzb. Acad. Wiss. Wien*, LXXVII.

extent deprived of oxygen, is conveyed to the pulmonary sinus. This is a large sinus running a nearly circular course round the base of the pulmonary chamber and receiving many small veins from the rectum. From the pulmonary sinus the blood is distributed through smaller

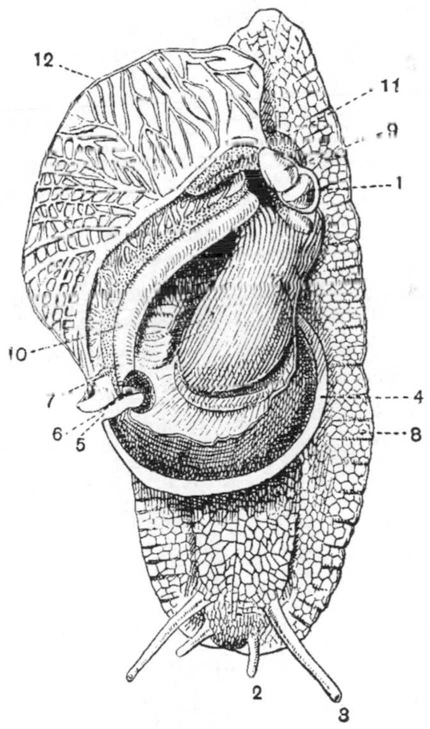

Fig. 42. *Helix pomatia*, with the upper wall of the pulmonary chamber cut open and folded back. From Hatschek and Cori.

1. Ventricle. 2. Anterior tentacles. 3. Eye tentacles. 4. Cut edge of mantle. 5. Respiratory pore. 6. Anus. 7. Opening of ureter. 8. Foot. 9. Auricle receiving pulmonary vein. 10. Rectum. 11. Nephridium. 12. Upper wall of pulmonary chamber.

vessels over the sides and roof of the pulmonary chamber which is in direct communication with the outer air. These afferent pulmonary vessels are connected by series of delicate vessels with corresponding efferent vessels, afferent and efferent occurring in regular alternation. The blood becomes aërated in passing through the delicate connecting vessels. The efferent pulmonary vessels gradually unite on the roof of the chamber to form the pulmonary vein which, as already mentioned, conveys the blood into the auricle of the heart. A portion of the blood has, however, a different course, passing into the glandular tissue of the nephridium and thence into the pulmonary vein without fully circulating within the pulmonary network of vessels. From this blood nitrogenous waste substances are eliminated by the nephridium.

The blood of *Helix* contains amœboid corpuscles in relatively small numbers and, dissolved in the plasma, an oxygen-carrier, hæmocyanin. This substance contains copper in combination with a proteid; when deoxidised it is colourless, but blue when oxidised. The blood of *Planorbis* contains hæmoglobin dissolved in the plasma. This substance is also found in the muscles of the pharynx and jaws of *Limnœa* and *Vivipara*[1]. The air within the pulmonary cavity is renewed chiefly by diffusion when the respiratory aperture is open. A more complete change of air is brought about when the snail is retracted into its shell by the columellar muscle; the resulting compression forces air out of the chamber; on the other hand

[1] Lankester, *Proc. Roy. Soc.* 1873.

when the animal expands again by the contraction of the muscular integument, of which the floor of the pulmonary chamber is a part, fresh air is drawn in. Many fluviatile Pulmonates admit water within the respiratory chamber, especially when living at considerable depths. It is probable that many of them are able to avail themselves of bubbles of oxygen given off by the leaves of aquatic plants[1]. There is no doubt that the skin is an important though subsidiary respiratory organ; this is especially the case among slugs and aquatic species. In many, *e.g. Limax flavus*, there are integumentary pigments with respiratory functions and having a marked affinity for oxygen.

Excretion. The nitrogenous waste substances are eliminated by the renal cells of the nephridium. Urates of ammonium and calcium are found in them and in the cavity of the nephridial sac. According to Cuénot[2] the uric concretions are discharged along the ureter at regular intervals of about a fortnight. It is probable that some excretory functions are performed by the walls of the pericardium, with whose cavity that of the nephridium communicates by the ciliated reno-pericardial canal.

Sense Organs. The sense of sight, or at any rate sensitiveness to light and shade, is possessed by all snails and slugs. In the terrestrial species eyes are situated at the summit of the upper pair of tentacles and can be withdrawn within the tentacle. The retraction consists of a "turning-outside-in" of the tentacle, the process beginning at the apex, so that the eye is the first

[1] Gain, *Journ. of Conch.* v.

[2] *C. R. Acad. Sci. Paris*, CXIX. 1894.

part to disappear within the shelter of the rest of the tentacle. *Limnæa* and the other aquatic pulmonate forms have a pair of eyes placed at the base of the single pair of solid contractile tentacles upon their inner side. The complex structure of the eyes would appear to indicate better visual power than has as yet been demonstrated by any experiments. There is strong reason to believe that the general surface of the integument is capable of appreciating differences of light and shade; and there is little if any evidence to prove that the eyes themselves are possessed of power superior to this.

It is rather by the sense of smell that terrestrial gastropods gain most information of the world around them. Moquin-Tandon[1] has recorded that one wet day he saw two slugs (*Limax maximus*) crawling from different points towards a decaying apple; he moved the fruit to the right, both slugs at once turned towards it, to the left, and again the slugs changed course making once more straight for the apple. When the tempting feast was held over them the tentacles and body were raised towards it. Many other observers have made similar experiments with various slugs and snails, all tending to show that the olfactory sense is extremely delicate and accurate. The precise position of the olfactory organs has given rise to much dispute. It seems probable that the specialised epithelia with which both upper and lower pairs of tentacles are provided at their apices are olfactory in function, and the results of Griffith's experiments[2] lend strong support

[1] *Mollusques de France*, I.

[2] *Proc. R. Soc. Edinb.* XIX. and No. 124, 1886–7.

to this view. The pedal gland[1] has by many been claimed as an olfactory organ: undoubtedly the character of its epithelial lining points to sensory functions of some sort, in addition to that of secreting mucus to which reference has already been made; it has not however been proved that the special sense here located is that of smell. Closely allied to smell is the sense of taste: this Simroth[2] considers to be diffused over the surface of the body with special developments under the lips and buccal membrane, in which situations special groups of sensory cells have been discovered by Smidt[3].

The otocysts have been assumed to be the auditory organs, but no evidence has been adduced to show that these animals are sensitive to those vibrations which produce in us the sensation of a sound. It is far more probable that these structures enable their possessor to determine the position of the body during locomotion, and to maintain its balance. Situated as the two otocysts are upon the ventral side of the pedal components of the subœsophageal ganglion mass, direct experiment upon them is impossible. Each cyst is a thin transparent sac about 0·15 mm. in diameter; it is lined internally with sensory and ciliated epithelial cells, and contains a fluid in which are suspended numerous minute calcareous bodies (otoconia); these are oval in shape and contain a "nucleus." During life the otoconia are maintained in constant motion by the

[1] Sochaczewer, *Zeit. wiss. Zool.* XXXV. and XXXVI. 1882; cf. André, *Rev. Suisse Zool.* 1894.

[2] *Zool. Anz.* 1882.

[3] Smidt, *Anat. Anz.* XVI.; cf. Gain, *Journ. of Conch.* VI.

vibrations of the cilia of the epithelial lining. The nerve supply is derived from the cerebral ganglia.

Reproduction[1]. All the pulmonate gastropods are hermaphrodite, and, as is usually the case when both male and female organs are present in the same individual, possess complicated reproductive systems. Inasmuch as the details of structure differ in various species we shall here confine ourselves to the arrangements found in *Helix pomatia*. The hermaphrodite gland or ovotestis situated in the upper coils of the digestive gland, is the seat of origin both of ova and spermatozoa; the latter are the first to mature, whence each individual functions first as male, and is termed "protandric." The ovi-sperm-duct (hermaphrodite duct) receives the spermatozoa or ova, as the case may be, from the gland and by means of its ciliated lining wafts them on to the confluence of the duct of the albumen gland. At the breeding season this gland increases in size and secretes a viscid albuminous substance with which the ova become surrounded. From the point of entry of the albumen duct onwards a separate course is provided for the ova and spermatozoa respectively. The spermatozoa pass, probably by their own vibratile movements, along the narrow and granulated portion of the "common duct"; the ova are urged by ciliary action down the wide and much folded section of the same duct. Thus the "common duct" though structurally one is functionally two tubes, a well-marked longitudinal constriction virtually separating the two internal channels.

[1] Garnault, *C. R. Acad. Sci.* Paris, CVI.; Baudelot, *Ann. Sci. Nat. Zool.* (4) XIX. 1863; Ashford, *Journ. of Conch.* IV. 1883.

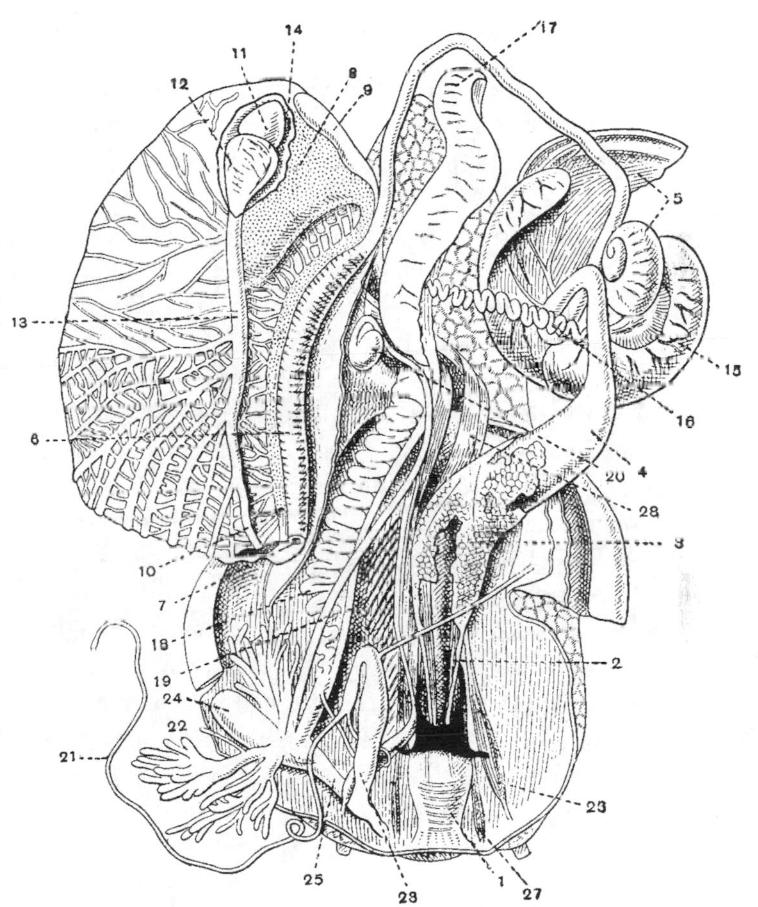

Fig. 43. *Helix pomatia*, opened and the viscera exposed. From Hatschek and Cori.

1. Pharynx. 2. Œsophagus. 3. Salivary glands, with duct. 4. Stomach. 5. Liver. 6. Rectum. 7. Anus. 8. Kidney. 9. Inflated commencement of ureter. 10. Opening of ureter to exterior. 11. Ventricle. 12. Auricle. 13. Pulmonary vein. 14. Opening of nephridium into pericardium. 15. Ovo-testis. 16. Common duct of ovo-testis. 17. Albumen gland. 18. Female duct. 19. Male duct. 20. Spermatheca. 21. Flagellum. 22. Accessory glands. 23. Penis. 24. Dart-sac. 25. Vagina. 26. Eye tentacle, retracted. 27. Anterior tentacle, retracted. 28. Muscles which retract the head, pharynx, tentacle, etc.

Further forward the two portions actually become distinct: the (male) vas deferens passes under the retractor muscle of the right tentacle to the muscular penis; the (female) oviduct proceeds nearly straight on to open into the vagina. Eventually both penis and vagina open at their common genital aperture upon the right side of the head.

Attached to these chief organs are numerous accessory structures, some belonging to the male, others to the female portion of the system. Springing from the junction of the vas deferens with the penis is the long whip-like flagellum: this, together with the hinder part of the penis secretes an elongated band of hardened mucus whose edges are rolled round to form a groove. The spermatophore, as this structure is called, receives into its groove masses of spermatozoa from the vas deferens and holds them as in a case. In the act of mating the spermatophore and its contents is transferred to the other snail *viâ* the penis introduced within the vagina of the mate: it is received into the duct of the receptaculum seminis (a long tube terminating posteriorly in a spherical head) which is given off from the posterior end of the vaginal cavity. In a few days the spermatophore is dissolved and the contained spermatozoa liberated in the receptaculum seminis. Here they remain until the ova of the impregnated snail are ready to be laid and are passing from the oviduct into the vagina. They then descend the duct of the receptaculum and fertilise the ova in the vagina. There are yet other organs, namely the mucous glands (multifid vesicles) and the singular dart-sac, connected with the vagina. The mucous glands secrete a highly refractive fluid which they discharge into the vagina; the fluid contains abundant calcareous matter and is thought to contribute to the outer covering of the eggs. The dart-sac is very muscular and contains the dart or *spiculum amoris*. This extraordinary weapon, which is flexible when within the sac but soon becomes brittle after removal, consists of a slightly curved, hollow, calcareous stem with four projecting and slightly twisted longitudinal flanges placed at right angles to each other; the stem and its flanges taper off to a fine point antoriorly. The muscular walls of the sac push the dart, with a slight screwing motion imparted by the flanges, into the body of the mate prior to sexual union. Owing to its slender attachment the dart frequently breaks off during use, and is left in the body of the mate. It appears to act as a sexual stimulant. A new dart is formed in about a week by the lining of the sac. The exact shape of the dart varies remarkably in the different species of *Helix*.

The eggs are laid during July and August in moist earth, under heaps of leaves and stones, and in similar places. They are about a quarter of an inch in diameter and enclosed in a tough calcareous shell containing a quantity of albumen and a small ovum. All the eggs of a season are not deposited simultaneously but in separate batches of from 20 to 60, one act of impregnation is however sufficient for the fertilisation of all[1].

Self-fertilisation is said to occur occasionally in the slug *Arion ater*[2] and in the aquatic snail *Limnæa auricularia*[3].

[1] Gaskoin, *Ann. Mag. Nat. Hist.* (2) IX.
[2] Wotton, *Journ. of Conch.* VII. 1893.
[3] Braun, *Nachr. Deutsch. Malak. Gesell.* XX.

It may be mentioned that in the *Limnæidæ* the male and female apertures are separate from one another, the latter being situated beneath the mantle flap some distance from the former which is placed as in *Helix*. The details of oviposition are better known in *Arion ater* than in *Helix pomatia*. Two individuals are recorded to have deposited respectively 396 and 477 eggs in the course of one autumn. Individual *A* deposited 80 eggs on October 10th, 110 on 16th, 77 on 25th, 82 on November 8th and 47 on 17th: *B* deposited 246 on October 13th and 14th, 9 on 26th, 121 on November 10th and 101 on 30th. The young hatched in from forty to seventy-four days, those of any one batch not hatching simultaneously. The young slugs were about 9 mm. long at birth; in five months they were 56 mm. long and were full grown in about a year and a half. Death occurred at about 2 years of age. In the case of *Helix aspersa* also details of the life-history have been published. The eggs hatched in August, about a month after being deposited; before the winter the shells had attained a diameter of ⅜ inch; by October of the following year the shells were 1 inch in diameter, but had no "lip"; in July of the next year a "lip" was formed, the diameter being now 1¼ inch (age, nearly 2 years). Mating took place in August of the year following (age, 3 years) and eggs were laid. The full term of life was about 5 years. The shell of an old snail can be distinguished by the smooth rolled back margin of the "lip." The length of life of the water snails *Limnæa* and *Paludina* is stated to be 3 to 4 years and 8 to 10 years respectively.

Enemies and Parasites. The enemies of our terrestrial

gastropods are most numerous and occur among nearly all classes of animals; in fact the *rôle* assigned to snails and slugs in the economy of the universe appears to be the altruistic one of affording food to others. Rats, field-mice, voles and hedgehogs prey upon various species of *Helix*. Shrews attack the smaller species and young individuals, while the water-shrew dives to hunt for *Limnœa* and *Planorbis*. Of birds the song-thrush is perhaps the most persistent enemy of *Helix*, whose shell the bird shatters by violent blows upon some favourite sacrificial stone. Blackbirds, starlings, ducks, geese, coots and landrails are all known to be devourers of slugs, and many smaller birds that frequent reed-beds and marshes feed freely upon the numerous small snails that abound in such localities. Frogs and toads are notoriously useful in clearing gardens of slugs and small snails, and are also partial, in common with newts, to various aquatic or semi-aquatic species. Trout and other fish are well known to thrive and grow most rapidly in streams and ponds with a dense gastropod population, both eggs and adults being equally acceptable. Over 350 shells of one species of *Valvata* have been taken from the stomach of a single eel.

Among invertebrate animals the Coleoptera provide enemies both to fluviatile and terrestrial species: the great water-beetle *Dyticus marginalis* is especially destructive to *Limnœidœ*, with an apparent preference for *L. stagnalis*; the *Silphidœ* destroy smaller land species and are stated by Cooke to break the shells by striking them against their own prothorax. The glow-worm, *Lampyris noctiluca*, and its allies, and the genera *Cychrus*, *Carabus* and the *Staphylinid* and *Hydrophilid* families of beetles all prey systematically upon snails.

Ants are recorded by Lowe as attacking *Helix aspersa*, and the larvæ of dipterous flies destroy large numbers of gastropod eggs. Ecto-parasites are relatively few, the best known instances being various species of mites (*Acari*). Of these *Conchophthirus* and *Philodromus* live upon the slime of various molluscs, while others inhabit the pulmonary chambers, as also does the Oligochæte Worm, *Chætogaster*. The Endoparasites are chiefly the immature Trematode and Cestode flatworms. An account of these will be found in a subsequent chapter (*vide* pp. 303—316). A Nematode, *Ascarioides limacis*, is said to occur in the eggs of *Limax* and in the same slug is found the Gregarine Protozoon *Isospora rara*.

Economic uses. Though generally harmful to agricultural industries some few gastropods are undoubtedly useful. All species of *Testacella* are predatory and clear the soil of numerous injurious Nematode worms, coleopterous and dipterous larvæ, and likewise of their own relatives, the herbivorous slugs and smaller snails.

As articles of human food snails have long been held in high esteem by many persons. The Romans imported choice varieties and fattened them in snail farms or Cochlearia, providing them with a diet of bran, flour, and herbs, with a judicious admixture of wine dregs. Pliny informs us that Fulvinus Hirpinus made the first enterprise in snail-farming at Tarquinium about the year 50 B.C.

Varro gives full directions for snail culture in his *De re rustica*, iii. 14: "You must choose," he says, "a suitable place out of doors and make a little moat all round the 'cochlearium,' or you will lose all your snails. The spot must not be exposed to the sun and should get heavy dews. If this is not at hand, you must make such a place as you find at the foot of rocks and mountains lapped by streams and lakes, and lay on your water-main and arrange the outflow so that it splashes around for some distance. Their feeding will be very little trouble to you. They live a long time, and when you want to get them ready for market you must throw in a few bay leaves and scatter some bran over them. And so let the cook cook them alive or dead, he generally does not know which."

Apicius, the author of a Roman cookery-book, supplies the following recipe:—"First catch your snails; wipe them over with a sponge; take off the lid [epiphragm] so that they may come out; put into their jar milk and salt for one day, for the remaining days milk only; keep them scrupulously clean by hourly removal of the excrement: when they have fed to such an extent that they cannot go back into their shells, fry them in oil."

At the present day snail-farms, or Escargotières, are maintained in various parts of Europe, nor are they entirely unknown in our own country. Darbishire describes one such establishment near Friburg in which some 60,000 to 80,000 *Helix pomatia* are fattened annually. The snails are collected by the country folk and brought to the wholesale farmer. By him they are kept for two or three months in a large meadow which is divided into small squares by hoardings about a foot high. These squares are filled with moss, and the enclosed snails fed upon a liberal diet of cabbage and lettuce leaves until they acquire a greenish-white colour. They are exported after the formation of the winter epiphragm and fetch about 17 francs per 1000. Paris alone is said to consume some fifty tons of snails daily during the time that they are in season.

In our own country, especially in the Western counties and in Yorkshire, *Helix aspersa* is in demand. In the neighbourhood of Bristol there exists a regular industry in collecting snails, which are highly esteemed by glass-blowers as "soothing to the chest." They are reputed also to be a cure for consumption, and the slime rubbed over the surface of a wart and allowed to dry is said to remove the excrescence. Lovell[1] states that they are employed in the manufacture of cream, and that a retired milkman pronounced it the most successful imitation known. The condition of the informant would appear to indicate that it was also remunerative.

[1] *Edible Molluscs*; cf. Cooke, *Camb. Nat. Hist.*, "*Molluscs and Brachiopods*," pp. 118—121.

BRITISH LAND AND FRESH-WATER GASTROPODS[1].

Class Gastropoda.

Sub-class ANISOPLEURA.

Order STREPTONEURA.

Sub-order PECTINIBRANCHIA. PROSOBRANCHIA.

Basommatophora. Eyes to inner side of cephalic tentacles, at their base; male and female generative apertures separate, on right side of neck. Aquatic.

Family VALVATIDÆ. Shell turbinate or discoidal; spire short and symmetrical, aperture round and entire; umbilicate; operculum horny; apex central. Aquatic.

genus *Valvata* Müll., *cristata* Müll., $\frac{1}{8}''$, whorls 5, shell a flat coil.

piscinalis (Müll.), $\frac{1}{4}''$, whorls 6, shell a broad cone, occasionally in brackish water.

Family VIVIPARIDÆ, genus *Vivipara* Montf. (*Paludina*). Shell turbinate with long symmetrical spire, whorls rounded; aperture oval; operculum horny and irregularly concentric. Aquatic.

contecta (Millet), $\frac{1}{2}''$, whorls 7, much rounded; suture deep, umbilicus deep.

vivipara (L.), $1\frac{1}{2}''$, whorls $6\frac{1}{2}$, rather convex; suture shallow, umbilicus very slight.

genus *Bithynia* Leach, eyes sessile; operculum calcareous. Aquatic.

leachii (Shepp.), $\frac{1}{4}''$, whorls 5, aperture nearly round.

tentaculata (L.), $\frac{1}{2}''$, whorls 6, aperture oval, angular above.

[1] The nomenclature here adopted is that given in the list of British Land and Fresh-Water Mollusca compiled by B. B. Woodward and published by the Conchological Society in October 1903. I am indebted to Messrs H. H. Brindley and F. C. Morgan of Cambridge for much kind help and information; and also to Gordon's "*Our Country's Shells,*" for many particulars relating to the shell. Undue prominence is really given to the shell, while the characters of the lingual ribbon (radula) and of the soft tissues have been deliberately withheld.

Family NERITIDÆ. Shell globular, thick; spire small; aperture semilunate, columellar side expanded; operculum calcareous with prominences on interior face, one of which articulates with columella.

genus *Neritina* Lam. *fluviatilis* (L.), $\frac{1}{3}''$, whorls 3, last more than two-thirds of whole shell. Aquatic; tolerates brackish water.

Family PALUDESTRINIDÆ. Shell conical, small, smooth, aperture entire; an umbilical cleft; eyes at base of tentacles. Aquatic (brackish).

genus *Paludestrina* Orbigny = *Hydrobia* Hartmn. Eyes on tubercles. Shell imperforate. Operculum horny.

confusa (Frauenfeld), $\frac{1}{7}''$, whorls 5 or 6. Suture deep; shell oval, stout; aperture oval; operculum spiral, not concentric.

jenkinsi (Smith), tentacles long, pointed; eyes on dark prominences.

taylori Smith.

ventrosa (Montagu), $\frac{1}{5}''$, whorls 6 or 7, suture rather deep; shell long and narrow; aperture oval.

stagnalis (Baster), whorls flattened, outer lip simple, inner reflected.

genus *Pseudamnicola* Paulucci. Shell perforated, small; aperture oval; operculum horny, spiral.

anatina (Drap.). [Identified by Boettger from specimens from Oulton lacking both animal and operculum.]

Family POMATIIDÆ. Shell spiral; spire elevated; aperture circular or nearly so; umbilicus small and narrow. Operculum calcareous, distinctly spiral.

genus *Pomatias* (Studer) = *Cyclostoma* (Montf.) *elegans* (Müll), $\frac{3}{5}''$, shell pale brown with purple tinge; whorls $4\frac{1}{2}''$; umbilicus twisted. Beneath stones, dead leaves, in hedgerows, etc. Not found in Scotland.

Family ASSEMANIIDÆ. Ocular tubercles long, contractile. Foot short, oval, truncate anteriorly. Shell as in Paludestrinidae. Operculum horny, spiral.

genus *Assemania* Leach. Spire short; aperture broad, oval, entire. Apex near internal edge of aperture.

grayana Leach. Estuary of Thames.

Family ACICULIDÆ. Shell small, cylindrical, thin, with long blunt spire; aperture oval; umbilicus straight; operculum horny; eyes sessile.

genus *Acicula* Hartman, *lineata* (Drap.), $\frac{1}{8}''$; outer lip of shell thin, inner lip spread. Beneath stones, moss, dead leaves, etc. Reaches southern counties of Scotland.

Sub-class ANISOPLEURA.

Order EUTHYNEURA.

Sub-order PULMONOBRANCHIA

Family LIMNÆIDÆ. Shell spiral, hood-shaped or coil-shaped, thin, horn-coloured; aperture without teeth, lip sharp; pulmonary sac protected by an external lobe; tentacles two, not retractile, eyes at base of tentacles.

sub-family *Ancylinæ*, shell hood-shaped; live attached like little Limpets to stones and weeds in the water.

genus *Ancylus* Geoffrey.

fluviatilis Müll., $\frac{1}{4}'' \times \frac{1}{3}''$, apex blunt, turning to right, aperture oval, genital and pulmonary apertures on left side.

genus *Acroloxus* Beck, *lacustris* (L.), $\frac{1}{10}'' \times \frac{1}{4}''$, apex sharp, turning to the left, aperture oblong, genital pulmonary apertures on right side.

sub-family *Limnæinæ*, shell spiral.

genus *Limnæa* Lam. em. Head large, tentacles triangular, shell thin, horny, usually dextral; aperture oval.

auricularia (L.), $1\frac{1}{8}''$, whorls 4 or 5, last enlarged; spire short, acute; inner lip forming small umbilical fissure behind it.

glabra (Müll.), $\frac{3}{5}''$, whorls 7 or 8, spire long, aperture with broad internal white rib.

involuta Harvey, $\frac{2}{5}''$, spire very short, sunken, suture deep, aperture pyriform. Found only near Killarney. The sole peculiar British species.

palustris (Müll.), 1'', whorls 6 or 7, enlarging gradually; spire long, suture deep.

peregra (Müll. em.), $\frac{3}{4}''$, whorls 5, last whorl enlarged, inner lip folded on pillar and forming a narrow groove.

stagnalis (L.), 2″, whorls 7 or 8, last enlarged suddenly, suture deep.

truncatula Müll., $\frac{2}{5}$″, whorls 5 or 6, spire long, suture very deep, umbilical fissure distinct.

genus *Amphipeplea* Nilsson, *glutinosa* (Müll), $\frac{1}{2}$″, spire very short, whorls 3 or 4, last forming nearly whole shell, suture shallow, aperture oval. Shell almost entirely covered by mantle, very thin, transparent.

Family PHYSIDÆ. Shell sinistral, lustrous, thin. Tentacles cylindrical, setaceous, not retractile. No operculum.

genus *Aplecta* Flem. em., *hypnorum* (L.), $\frac{3}{4}$″, shell spiral, oval, very convex, apex blunt; umbilicus a narrow cleft; aperture simple, sharp-edged. Mantle devoid of finger-like projections over the shell. Whorls 6 or 7, spire long.

genus *Physa* Drap., *fontinalis* (L.), $\frac{1}{2}$″, shell spiral, oval, oblong, transparent; aperture vertical, elongate, simple, sharp-edged. Mantle with finger-like prolongations over the shell, whorls 4 or 5, spire short.

The species *Ph. acuta* Drap. and *Ph. heterostropha* Say. are introduced.

Sub-family *Planorbinæ*.

genus *Planorbis* Geoff., shell a flat coil, dextral; aperture crescentic; tentacles cylindrical.

fontanus (Lightfoot), $\frac{1}{6}$″, whorls 4 or 5; thin, shining, compressed, keeled in middle, last whorl not embracing the rest.

(*a*) whorls rounded.

albus Müll., $\frac{1}{4}$″, whorls 5; grayish white, with longitudinal striæ, last whorl large.

contortus (L.), $\frac{1}{5}$″ whorls 8, compact, shell thick; aperture semilunar; umbilicus large, deep.

corneus (L.), 1″, whorls 5 or 6, rounded above and below; suture deep; aperture narrow, crescentic, umbilicus large, deep.

glaber Jeff., $\frac{1}{7}$″, whorls 5, last whorl large, shell convex above, concave below, smooth, no circular striæ; umbilicus large.

(*b*) whorls keeled or angulated.

carinatus Müll., $\frac{1}{2}$″, whorls 5 or 6; keel in middle of whorls, last whorl large, aperture obliquely oval.

umbilicatus Müll., $\frac{3}{4}$″ whorls 5 or 6; keel on lower side of whorls, last whorl large; aperture rhombic.

dilatatus Gould, $\frac{1}{10}''$, whorls 2, angulated, aperture very large and squarish, umbilicus small, deep : found only in canals near Manchester ; probably introduced by shipping.

crista (L.), $\frac{1}{10}''$, whorls 3, last whorl rather large, small imbricated ridge, aperture oval, umbilicus large.

spirorbis (L.), $\frac{1}{4}''$, whorls 5 or 6, one side concave, the other flat ; keel frequently blunt, aperture roundish, often ribbed with white, umbilicus wide and shallow.

vortex (L.), $\frac{3}{8}''$, whorls 6 to 8, upper side concave, under side flat, keel on lower margin and always well marked ; aperture rhombic ; umbilicus large, shallow.

genus *Segmentina* Fleming, shell flattened, last whorl embracing the rest, keeled ; aperture transverse, angular ; interior with lamellar, teeth grouped in threes.

nitida (Müll.) = *Planorbis lineatus* (Walker), $\frac{1}{6}''$, whorls 4.

Family AURICULIDÆ. Shell spiral, spire short, body whorl large, aperture toothed, tentacles two, eyes at base.

genus *Caryohium* Müll., tentacles relatively large, cylindrical ; eyes below and behind base of tentacles.

minimum Müll., $\frac{1}{15}''$, shell transparent, oblong, finely striated transversely, aperture oval, toothed, margin thickened ; whorls $5\frac{1}{2}$. In damp places, under rotten wood and dead leaves.

genus *Phytia* Gray, tentacles cylindrical, swollen at apex ; foot not divided.

myosotis (Drap.).

genus *Ovatella* Bivona, tentacles short, compressed ; foot divided by a transverse groove.

bidentata (Montagu).

Sub-order PULMONATA.

(*Stylommatophora.* Eyes on summit of two hollow tentacles. Almost exclusively terrestrial.)

Family ZONITIDÆ.

genus *Vitrina* Drap., *pellucida* (Müll.), $\frac{1}{4}''$, shell greenish, subglobular, thin, vitreous, whorls few, increasing rapidly, no umbilicus, aperture large, ear-shaped, animal incapable of retreating entirely within shell. In shady and moist places, under stones and moss.

genus *Vitrea*, Fitzinger, shell orbicular, depressed, thin, whorls increasing regularly, aperture oblique, umbilicus more or less distinct. Animal capable of retreating entirely within shell. Most occur under stones or wood in damp places, wet moss; avoid the light.

lucida (Drap.), $\frac{1}{2}''$, whorls 6 or 7, transversely striated, fawn above, bluish-white below; aperture oval, very oblique, umbilicus wide; animal cobalt blue.

pura (Alder), $\frac{1}{10}''$, whorls 4, last=half of shell; striæ circular, umbilicus narrow, deep.

radiatula (Alder), $\frac{1}{8}''$, whorls 4 or $4\frac{1}{2}$, striæ from whorl to whorl not interrupted by sutures; aperture semilunar, oblique; umbilicus moderately deep.

nitidula (Drap.), $\frac{3}{8}''$, whorls 4 or 5; striæ interrupted by sutures; spire slightly produced; aperture oblique, small.

crystallina (Müll.), $\frac{1}{16}''$, whorls $4\frac{1}{2}$ to 5, body whorl same as preceding whorl, aperture crescentic; umbilicus small.

alliaria (Miller), $\frac{1}{4}''$, whorls 5, last moderately large, exposing second whorl; aperture semilunar, narrow.

rogersi B. B. Woodward, $\frac{1}{4}''$, whorls 5 or $5\frac{1}{2}$, body whorl occupying half of shell; shell thin, glossy, compressed, transversely striated, dark horn-coloured; aperture $\frac{3}{4}$ of a circle; umbilicus narrow, deep.

cellaria (Müll.), $3\frac{1}{4}''$, whorls 5 or 6, last large; shell pale horn.

genus, *Zonitodes* Lehm., shell striated above, whorls increasing gradually.

nitidus (Müll.), $\frac{1}{4}''$, whorls 5, shell sub-globose and brownish, aperture oblique and roundish; umbilicus narrow and deep.

excavatus (Bean), $\frac{1}{4}''$, whorls $5\frac{1}{2}$; shell thin, horn-coloured; aperture small, crescentic, umbilicus large, disclosing all the whorls.

genus *Euconulus* Reinh., shell thin, perforate.

fulvus (Müll.), $\frac{1}{10}''$, whorls $5\frac{1}{2}$ or 6; shell keeled, conical, coloured, aperture crescentic, umbilicus large, exposing second whorl. In vaults and wells.

Family HELICIDÆ. Animal terrestrial; shell spiral, globose or coil-shaped; tentacles four, invaginable; eyes at tip of upper longer pair.

genus *Helix* L., shell globular or flattened, aperture circular or oval, outer lip generally thick and without a rib, sometimes reflected and toothed. Most species to be found in woods, hedges, gardens, etc., especially in moister parts.

(1) without umbilicus, *i.e.* concealed.

nemoralis L., 1″, lips brown, common everywhere.

aspersa Müll., $1\frac{3}{5}$″, lips white, whorls $4\frac{1}{2}$; spiral bands generally confluent, crossed by transverse patches. Not common in limestone districts.

hortensis Müll., 1″, lips white, whorls $5\frac{1}{2}$, spiral bands distinct and not crossed by transverse patches.

(2) with umbilicus.

(*a*) with labial rib.

* *pisana* Müll., $\frac{3}{4}$″, shell without keel, striped and mottled; umbilicus small and oblique.

(*Hygromia* Risso.) *rufescens* (Penn), $\frac{1}{2}$″, shell bluntly keeled; reddish-brown, frequently with white band on last whorl.

hispida (L.), $\frac{1}{3}$″, shell not keeled, not striped, may be devoid of labial rib.

granulata (Alder), $\frac{1}{3}$″, shell not keeled, not striped, greyish-white, downy and granulated; umbilicus very small.

(*Helicella* Fér.) *virgata* (Da Costa), $\frac{1}{2}$″, shell without keel, striped but not mottled; umbilicus narrow and deep. Dry downs with short turf.

cantiana (Mont.), $\frac{3}{4}$″, shell without keel, unstriped, tinged with rose, umbilicus narrow and deep.

* *cartusiana* (Müll.), $\frac{1}{2}$″, shell without keel, unstriped, horn colour, umbilicus minute.

caperata (Mont.), $\frac{1}{3}$″, shell keeled and striped with coloured bands. Dry downs with short turf.

(*b*) without labial rib.

* *Helix pomatia* L., $1\frac{3}{4}$″, shell globose, striated, pale yellow with broad light brown bands.

(*Helicigona* (Fér.) Risso.) *arbustorum* (L.), $\frac{4}{5}$″, shell globose, striated, brown with blackish band along the middle of each whorl.

lapicida (L.), $\frac{2}{3}$″, shell depressed, rufous brown, aperture oval, angulated, umbilicus large. Dry downs with short turf.

* Occur only in southern and western counties.

(*Helicella* Fér.) *itala* (L.), $\frac{5}{8}''$, shell depressed, greyish with brown bands, aperture nearly circular, umbilicus large. Dry downs with short turfs.

* (*Helicodonta* (Fér.) Risso.) *obvoluta* (Müll.), $\frac{1}{2}''$, shell depressed, rufous brown, aperture triangular and toothed, umbilicus large.

(*Vallonia* Risso.) *pulchella* (Müll.), $\frac{1}{10}''$, shell milky-white.

(*Hygromia*) *fusca* (Mont.), $\frac{1}{3}''$, shell wrinkled.

* *revelata* (Fér.), $\frac{1}{3}''$, greenish, bristly, transparent; umbilicus large.

(*Acanthinula* Beck.) *aculeata* (Müll.), $\frac{1}{7}''$, periostracum raised, whorls 4 to $4\frac{1}{2}$, aperture semicircular.

lamellata (Jeff.), $\frac{1}{7}''$, periostracum raised, whorls 6, aperture crescentic.

(*Cochlicella* (Fér.) Risso.) *barbara* (L.) = *acuta* (Müll.), $\frac{3}{8}''$, whorls 8 or 9; shell long and spiral, aperture oval, umbilicus small.

Endodontidæ Pilsbry.

(*Punctum* Morse.) *pygmæa* (Drap.), $\frac{1}{16}''$, shell pale horn-coloured.

(*Sphyradium* Charp.) *edentulum* (Drap.), $\frac{1}{10}''$ = *vertigo edentula*, shell faintly striated.

(*Pyramidula* Fitz.) *rupestris* (Drap.), $\frac{1}{9}''$, shell blackish-brown.

rotundata (Müll.), $\frac{1}{4}''$, shell bluntly keeled.

An enormous number of varieties have been described, especially of *nemoralis*, *hortensis*, and *arbustorum*.

Family SUCCINEIDÆ. Shell oval or oblong, thin; spire short, last whorl large, mouth obliquely oval; tentacles four, retractile. Aquatic or semi-aquatic, on stalks and leaves of water plants, generally above the surface.

genus *Succinea* Drap.

oblonga Drap., $\frac{1}{4}''$, light horn-coloured, whorls 4, suture oblique, very deep.

elegans Risso, $\frac{3}{5}''$, oblong amber-coloured, or greenish-yellow, whorls 3 or 4, suture very oblique but not deep.

putris (L.), $\frac{3}{5}''$, oval, whorls 3 or 4, mouth oval and $\frac{2}{3}$ length of shell, suture rather oblique and very deep.

Family STENOGYRIDÆ. Shell long, cylindrical, thin, translucent; spire long, whorls rapidly increasing in size, aperture notched.

* Occur only in southern and western counties.

genus *Cæcilioides* (Hermans), *acicula* (Müll.), $\frac{1}{6}'' \times \frac{1}{20}''$, $5\frac{1}{2}$ whorls, compressed and drawn out. Roots of trees and under stones; usually buried; never crawls on surface.

genus *Cochlicopa* (Fér.) Risso, whorls from 6 to 8, *lubrica* (Müll.), $\frac{1}{4}''$, aperture of shell without teeth or folds.

genus *Azeca* (Leach), aperture pyriform with teeth or folds upon its rim, *tridens* (Pult.). In moss, decayed leaves, etc., at foot of hedgerows.

Family ENIDÆ.

genus *Ena* (Leach), shell long and spiral, aperture oval, umbilicus small.

montana (Drap.), $\frac{5}{8}''$, whorls $6\frac{1}{2}$ or $7\frac{1}{2}$, umbilicus oblique and deep, S. and W. counties only.

obscura (Müll.), $\frac{1}{3}''$, whorls $6\frac{1}{2}$, umbilicus narrow, not deep; shell golden-brown, thin and translucent.

Family VERTIGINIDÆ (PUPIDÆ). Shell cylindrical or oblong, occasionally furnished with internal lamellæ; whorls numerous, narrow, last whorl no broader than the preceding; aperture generally with one or more teeth.

genus *Jaminia* Risso, *Pupa* (Drap.), aperture of shell oval or lunate. In moss on stones, tree branches, etc.

muscorum (L.), $\frac{3}{5}''$, aperture with one tooth, white rib behind outer lip, suture deep.

cylindracea (Da Costa), $\frac{1}{25}''$, aperture with 1 tooth, suture shallow and oblique.

anglica (Fér.), $\frac{1}{7}''$, aperture with 5 folds.

secale Drap., $\frac{1}{3}''$, aperture with 8 or 9 teeth, 2 or 3 on pillar, 2 on pillar lip, 4 inside outer lip.

genus *Vertigo* Müll., aperture of shell more or less angular. On old walls, under stones and among grass roots.

(1) SHELL SINISTRAL.

angustior Jeff., $\frac{1}{16}''$, teeth 4 or 5, aperture triangular.

pusilla Müll., $\frac{1}{14}''$, teeth 6 or 7, aperture square.

(2) SHELL DEXTRAL.

(*a*) No teeth.

minutissima (Hartmn.), $\frac{1}{14}''$, striations distinct.

(*b*) Teeth 4 or 5.

moulinsiana (Dupuy), $\frac{1}{12}''$, shell with rib, umbilicus open. A restricted species; has been found at Wicken Fen, Cambs. (H. H. B.)

pygmæa (Drap.), $\frac{1}{14}''$, rib very stout, umbilicus narrow.

alpestris (Alder), $\frac{1}{14}''$, shell not ribbed, suture very deep.

(*c*) Teeth 6.

substriata (Jeff.), $\frac{1}{15}''$, aperture semi-oval; *tumida* $\frac{1}{16}''$, aperture heart-shaped.

antivertigo (Drap.), $\frac{1}{13}''$, aperture triangular.

Family CLAUSILIIDÆ.

genus *Balea* (Prideaux) Gray, shell sinistral, aperture ovate. Bark of trees and mossy walls.

perversa (L.), $\frac{1}{4}''$; top whorl shining.

genus *Clausilia* Drap., shell sinistral, aperture with *clausilium*. Bark of trees and under stones.

(1) margin of clausilium notched.

biplicata (Mont.), $\frac{5}{8}''$, 2 folds on pillar, shell streaked.

laminata (Mont.), $\frac{3}{4}''$, 3 or 4 folds on pillar, shell smooth.

(2) margin of clausilium entire.

rolphii Leach, $\frac{1}{2}''$, 4 or 5 folds on pillar; shell fusiform and streaked.

bidentata (Ström.), $\frac{1}{2}''$, 3 folds on pillar.

Family LIMACIDÆ. Animal terrestrial; shell internal, beneath the mantle, rudimentary or shield-shaped; tentacles four, cylindrical, invaginable; eyes at tip of dorsal pair.

genus *Milax* Gray, respiratory aperture in posterior half of shield, which is shagreened; a well-marked dorsal keel along body.

gagates (Drap.), colour dark lead, sides pale.

sowerbyi (Fér.), colour tawny yellow, keel yellow.

genus *Limax* L., respiratory aperture in posterior half of shield, shield concentrically wrinkled.

flavus L., animal 4″; yellowish with black-brown spots, but variable; head and tentacles grey; shell $\frac{1}{3}''$ long $\times$ $\frac{1}{4}''$ broad.

arborum Bouch-Chant, 3″, grey with yellow spots, dark central and lateral stripes; shell nearly flat.

maximus L., 6″, grey with black spots but varies much, tentacles long, purple; shell oblong, $\frac{1}{2}'' \times \frac{1}{3}''$. Var. *cinereo-niger* (Wolf), 4″, respiratory aperture with margins darker than rest of body; sole of foot with white median band.

genus *Agriolimax* (Mörch.), *agrestis* (L.), animal $1\frac{1}{2}''$; shield marked with concentric lines, colour very various; shell small, oblique.

lævis (Müll.), $\frac{3}{4}''$, dark violet brown; shell square, minute; respiratory aperture near centre of right margin of mantle.

Family ARIONIDÆ Gray. Shield covering all the surface of the body.

genus *Arion* Fér., respiratory aperture half-way along shield, shell of loose calcareous grains, covered by hinder part of shield.

ater (L.), 4″, shell of loose calcareous granules; tentacles black.

fasciatus Nilsn., $\frac{1}{4}''$, whitish-grey with lateral stripes.

hortensis Fér., $\frac{1}{2}''$, shell granular, in an oval mass; animal with grey stripes.

subfuscus (Drap.), $\frac{1}{4}''$, shell of isolated calcareous granules; animal reddish with two black stripes.

elongatus Collinge.

intermedius Normand, $\frac{3}{4}''$, light grey, tinged with yellow; body glistening.

genus *Geomalacus* Allman, respiratory aperture near front edge of shield, shell unguiform, leathery, imbedded in shield.

maculosus Allman, 2″, black, spotted with yellow; foot brown, sole light yellow, occurs only in S.W. Ireland; probably introduced from Spain.

Family TESTACELLIDÆ Gray. Animal terrestrial, slug-like; shell rudimentary, ear-shaped and external, carried at the posterior end of body; tentacles four, retractile, eyes at tip of upper pair. (Probably not truly indigenous, but now well established.)

genus *Testacella* Cuvier.

haliotidea Drap., shell $\frac{1}{3}'' \times \frac{1}{5}''$, ear-shaped, convex; animal whitish or cream-coloured; length $3\frac{1}{2}''$—4″.

scutulum Sby., shell $\frac{1}{3}'' \times \frac{1}{6}''$; ear-shaped, flat or concave; animal tawny yellow with brown spots, length about 3″.

maugei Fér.; shell $\frac{3}{5}'' \times \frac{1}{4}''$, convex, semicylindrical; animal narrow in front, swollen behind, colour grayish-white to black-brown, spotted; length $2\frac{1}{2}''$—$3\frac{3}{4}''$.

CHAPTER VIII.

FROGS, TOADS, AND NEWTS.

THERE is no vertebrate animal whose anatomy has been more thoroughly investigated and described than the Common Frog, *Rana temporaria*. The specific name refers to the large black patch on the temporal region of the head. The general colour of the animal, and especially of its dorsal surface, is liable to great variation, not only in different individuals, but in the same individual on different occasions. The most usual coloration is perhaps a mottled green and brown, but all conditions between yellowish-red and dark brown-black are of frequent occurrence. The markings upon the hind legs are chiefly transverse. The ventral surface is yellow, pale or bright, and is sometimes spotted. The colours are due to stellate pigment-cells, that occur sparsely in the epidermis but abundantly in the loose layer of subcutaneous connective tissue. Nerve fibres pass into direct connexion with these cells, and by appropriate impulses regulate their shape, causing them to contract to mere pin-points or to spread out over a relatively large area according as it is necessary to minimise or display to utmost advantage any particular pigment. The action is almost certainly a reflex depending upon sensory impulses received by the retina of the eye. It is this power which enables a frog to imitate with very fair success the general colour effect of the varied surroundings among which it passes its life. As it squats on the ground among the grass it may well escape notice until it jumps. In this position the value of the pale ventral edges of the flanks becomes evident in lessening the effect of the shadow cast by the body of the animal, and thus causing it to appear not to project above the surface of the ground.

The surface of the skin is smooth and moist. The outer horny layer is, except on the back and on the under side of the toes, very thin. This fact may be realised at the periodic casting or "sloughing" of the horny layer, when it appears as a translucent pellicle which the frog, economically, swallows. Beneath the surface layers are numerous goblet- or mucous-cells, which are stated by Pfitzner[1] to secrete a substance which separates the upper layer from that below, and brings about a complete shedding of the skin. Two kinds of glands are present in the skin, serous and mucous: the former are found chiefly upon the back and correspond to the poison-glands of the skin of toads; the latter are more numerous and more evenly distributed over the body, their protoplasm swells up freely on the addition of water. The ducts of both kinds of gland open upon the surface, and play an important part in keeping the skin moist and permeable by the gases of respiration.

[1] *Morph. Jahrb.* VI.

The head of the frog is bluntly triangular. From its otherwise flat dorsal surface the eyes project for some distance, and the nostrils but slightly, above the general level. The animal is thus enabled both to breathe and to see above the water while keeping the rest of its head and body completely hidden beneath the surface. The upper eyelid is attached to the eyeball and moves with it. The lower eyelid (nictitating membrane) alone is capable of winking, in which act it is raised as a transparent curtain over the eyeball by a special muscle, while the eyeball itself is retracted. When frogs are beneath the surface of water, and especially when hibernating in this situation, the lower eyelid is frequently kept half or completely drawn across the front of the eyeball. The eye sockets have no bony floor but are separated from the mouth cavity by soft tissues only. It is thus possible for the eyes to be drawn down into the mouth. The muscle that brings this movement about is known as the *retractor bulbi*: it arises from the parasphenoid bone and forms a fairly complete sheath surrounding the hinder hemisphere of the eyeball. The tendons of the nictitating membrane are attached to the *retractor bulbi* in such a way that when the muscle contracts the membrane of necessity rises across the front of the cornea. Elevation of the eyeball is caused by a muscle, the *levator bulbi*, which arises from the fronto-parietal, sphenethmoid and other bones, and passes obliquely backwards and outwards as a sheet below the eyeball to be inserted into the upper border of the maxilla. A few fibres pass from the muscle into the lower eyelid and serve to pull it down again as the *levator* pushes the eyeball up.

The tympanic membrane of the ear is situated a short distance behind the eye in the black patch already referred to: its margin is marked by an almost circular ring of cartilage which can be detected through the skin; the disc thus circumscribed is slightly concave and has attached to its centre internally the rod-like *columella* establishing connexion with the internal ear.

There is no neck. The trunk, when the animal is squatting, presents a very decided dorsal hump about halfway along its length. This projection is caused by the long hip-bones (ilia) which are at this point attached to the last (ninth or sacral) vertebra, and thus transfer the movements of the hind legs to the trunk and entire body.

The disparity of front and hind limbs is very striking. The former are short and of quite secondary importance in locomotion either on land or in the water: they are used in slow or scrambling movements through herbage and also to raise the fore-part of the body and head and impart an upward inclination to the animal when it jumps, and thereby increase its trajectory. In alighting the front feet are the first to reach the ground, and act as spring buffers for the protection of the head and body against the shock of impact. When at rest the animal keeps the elbows turned outward and the four digits of each front foot turned in towards the middle line.

The hind legs are relatively enormous, being more than twice as long as the front legs. The thigh and shin together are about double the length of their serial homologues the upper arm and fore-arm, but the ankle and foot are between three and four times as long as the wrist and hand. Most remarkable is the great development of the astragalus and calcaneum, tarsal bones which in the great majority of animals are short and compact: here they are lengthened out to such an extent as to render it possible for a beginner to mistake these bones for tibia and fibula. The toes are also of very great length, especially the 4th; the hallux, or "big-toe," is decidedly the shortest, and at its base is the vestige of yet another digit. The leg may be regarded as composed of three levers folded, in the resting attitude, on one another; when a leap is taken the three levers are straightened out so as to lie in virtually one and the same straight line, the toes giving the final push off from the ground. The hind leg is the main instrument in swimming: the feet are webbed as far as the penultimate phalanx of the 5th, 4th, and 3rd digit on its outer side: there is a moderate web between the 3rd and 2nd, and between the 2nd and 1st digits. Thus the outer and upper portion of the foot makes the stronger part of the stroke in swimming —a fact that is of no small importance in enabling the frog to dive quickly below the surface on entering the water. It must be borne in mind that the adult frog, except at the breeding season, prefers to be on land near the water rather than in the water itself. He hastily takes refuge in the water when alarmed and vanishes beneath the surface. It is therefore of vital importance that the web of the hind foot should urge the body downward. The front feet are not webbed, and take but little part in actual swimming, but are usually spread out to help in supporting the animal when resting in the water: they come into active use as it crawls up the bank on to the land again. When the animal is swimming slowly the front limbs do make feeble strokes. During rapid swimming however they are kept closely pressed to the sides of the body and extended straight backwards. The common toad on the other hand generally swims holding the front limbs flexed at the elbow with the hands beside the head, and the digits pointing forward. If severely pressed he will dive and throw the front limbs into the attitude adopted by the frog. But the natterjack toad, which is a more thoroughly terrestrial animal, cannot by any provocation be induced to put the front limbs back against the sides of the body. With comparatively weak strokes of the hind legs this animal paddles along on or very near to the surface of the water with the hands in contact with the sides of the face.

Food and Digestion. The food consists of snails, slugs, beetles, earwigs, dipterous flies, worms, and other small invertebrate animals. The prey is seized by means of the tongue. This is a bluntly bilobed muscular organ attached anteriorly to the floor of the mouth, its bifid free end being directed down the throat. In taking food the tongue is flicked out of the mouth and back again with extraordinary rapidity: its free end is thrown by a somersault an inch or more in advance of the tip of the jaw, and whipped back again to the pharynx, carrying the prey with it. As the tongue is being projected it wipes from the roof of the mouth the secretion of the intermaxillary glands—a mass of convoluted tubes between the premaxillæ and opening on the front part of the palate. This secretion is extremely adhesive mucus, and forms a very effective instrument in capturing the prey. This remarkable movement of the tongue is the result partly of direct muscular action and partly of injection of fluid into the lingual lymphatics. According to Hartog[1] the action is commenced by the petrohyoid and geniohyoid muscles lifting the hyoid cartilage and bringing it forward; the genio-glossal muscle, running from the chin into the tongue, and stylo-glossal muscle begin the dilation by shortening and widening the cavity of the tongue, but the chief muscle concerned in projection is the mylohyoid, which, running across the floor of the mouth, by forcible contraction drives lymph into the tongue. In retraction the hyoglossus, from the bony part of the head to the tongue, and the intrinsic lingual muscles draw the tongue back into the mouth; the sterno- and omo-hyoids, passing back to the shoulder girdle, pull back the attachments of the tongue to the hyoid cartilage, and closure of the mouth forces the tongue against the palate, and presses out the rest of the lymph.

A small animal, such as a beetle, when seized by the tongue is thus conveyed straight to the top of the gullet and swallowed without being touched by the teeth. Larger animals such as worms and slugs are seized between the jaws and swallowed more gradually. It is in dealing with relatively large and slippery prey of this character that the teeth are of importance. They are carried upon the premaxillæ and maxillæ, bordering the upper jaw, and in two patches upon the vomers in the roof of the mouth; the mandible, or lower jaw, is edentulous. All the teeth slope inwards. Their function is clearly that of holding

[1] *C. R. Acad. Sci.* Paris, t. 132, and *Ann. Mag. Nat. Hist.* (9) VII.

the prey and preventing it from slipping out of the mouth; they cannot be said to bite, either in the sense of cutting or of grinding. The eye-balls however are very important organs in gripping the prey in the mouth. At each gulp made as a worm is being swallowed both eyeballs are forcibly depressed into the mouth cavity, thrusting the worm against the floor of the mouth and downwards into the pharynx. If the worm chances to be decidedly upon one side of the mouth the eyeball of that side alone is depressed, the other remaining prominent. The toads make use of their eyeballs in the same manner.

There is no sharp line of demarcation between the mouth, pharynx, and œsophagus (gullet), but the commencement of the stomach is indicated by a decided curve of the alimentary canal away from the median line towards the left side of the body. The epithelium lining the mouth, œsophagus, and parts of the stomach is ciliated; that of the rest of the stomach is highly glandular; no less than four different types of gland have been described by various authorities. The stomach lies almost parallel with the long axis of the body, but turns slightly towards the right. At its extreme right-hand end the circular muscle coat of its wall is thickened to form the pylorus—a ring of muscle which acts as a guard at the junction of the stomach with the small intestine and permits the passage of food in a fine state of division only. The gastric juice is decidedly acid and effects the solution of the shells of snails as well as the digestion of proteid food substances. It seems probable that the chitinous exoskeleton of insects is to a large extent dissolved; it is undoubtedly broken up and much of it disappears, but the details of the process are not known. The mucous membrane of about the first half of the small intestine is thrown into two series of transverse folds resembling the semilunar valves of the human heart; they are so arranged as to prevent reflux of the food towards the stomach while permitting it to move on towards the large intestine. These folds are replaced by irregular wrinklings in the portion of the duodenum immediately adjoining the pylorus. In the last half of the small, and throughout the large, intestine the mucous membrane is thrown into longitudinal folds whose appearance differs with the state of distension or contraction of the organs concerned. By these folds and also by the villi the area of the internal absorbing surface of the intestines is largely increased. The length of the alimentary canal is relatively small: this is in accordance with the animal nature of the food: in the tadpole, which is chiefly, if not entirely, herbivorous, the length of the canal is far greater in proportion.

Liver and Pancreas. The liver lies in the anterior part of the abdomen, close behind the heart, and covers the stomach ventrally. It is divided into several lobes, and between the two main lobes, right and left, is situated the gall-bladder, with which some of the ducts of the liver are connected. The common bile-duct, by which the green-coloured bile is conducted away, passes through the pancreas which lies in the loop between the stomach and duodenum: about halfway along the pancreas it receives the pancreatic ducts, and eventually enters the duodenum very obliquely with a slit-like aperture. Thus the bile and pancreatic juice are poured simultaneously upon the food almost immediately after it issues from the stomach. By these two fluids the digestion of the proteids, carbohydrates and fats of the food are completed. The green colouring matter of the bile is, in part at any rate, a derivative from the hæmoglobin, or red colouring matter of the blood. Glycogen is stored up in the hepatic cells as the result of digestion and absorption in the intestine: the absorbed food is conveyed to the liver by the portal vein and there converted into glycogen, pending the demands of the general tissues of the body. As occasion arises it is converted into more soluble material, a sugar, and sent into the main blood-stream *viâ* the hepatic veins and inferior vena cava. Fat globules are also contained by the liver cells. The storage function of the liver is one of considerable importance, especially during hibernation and at the breeding season; the weight of the organ exhibits a well-marked seasonal variation in accordance with the amount of reserve food contained. The details of this phenomenon have been worked out by Alice Gaule[1] in *Rana esculenta.* The breeding season of this frog is in May, June, and July. The table on the next page shews the average weight of the liver in the two sexes month by month.

It will be observed that the liver is most depleted in both sexes in June, the middle of the breeding season, and that it reaches its maximum weight in September when the system has recovered from the exhaustion of spawning. Throughout the winter the reserves are being steadily used up, with no recovery by the female, the average

[1] *Pflüger's Archiv f. ges. Physiol.* LXXXIV.

Month	Weight of male liver	Weight of female liver
January	10 grms.	13 grms.
February	10	12·5
March	13	11
April	10	10
May	12	9
June	5·5	7·5
July	7·5	11
August	6	12
September	22·5	27
October	18	25
November	22	25
December	18	22

weight of whose liver is greater than that of the male, but with a slight recovery in March and in May by the male. It is probable that this general difference depends upon the fact that the ovaries of the female make a great and continuous demand upon her system throughout the whole period of maturation, so that in spite of renewed feeding in the Spring there is no recuperation in the liver. In the male, however, there is no such continuous drain but rather a sudden call upon the reserves at the actual time of pairing—a call due not only to the discharge of the spermatozoa but also to the muscular exertions of the male at that season. This call is marked in vigour by the sudden reduction of the liver to rather less than half its weight in June.

Reserves of food are also laid up in the fat-bodies. These have no direct connexion with the digestive system, but may conveniently be dealt with here. They are bright yellow, finger-like bodies grouped in front of the testes or ovaries as the sex may be. They develop from the anterior portion of the genital ridges whose posterior portions alone give rise to the sexual organs[1]. In the autumn they are of great size and loaded with fat-cells, a certain amount of lymphatic tissue being also present. In the spring they are much reduced. It is probable that they also perform other functions at all seasons of the year, but on this point we have no precise knowledge.

Circulation. The blood consists of a clear, almost colourless fluid, the plasma, in which float corpuscles of two sorts—'red' and colourless. The former are by far the more numerous, though the exact proportion is subject to considerable fluctuation. The 'red' corpuscles are oval discs, about 0·02 mm. long by 0·01 mm. wide, containing an oval nucleus which causes the central region of the disc to bulge outwards on each face. Their colour is yellow rather than red, though when any thickness of them is seen they produce a red effect. The colouring matter is hæmoglobin; it acts as a carrier of oxygen from the respiratory organs to the general tissues of the body. The colourless corpuscles are of several different varieties and sizes, but are smaller than the 'red'; they possess the power of independent movement like an *Amœba* and contain one or more nuclei. They can also migrate through the walls of the blood vessels and are found in nearly all tissues of the body, but especially in the connective tissues. They probably play an important part in the removal and repair of effete or injured tissues, and

[1] Marshall and Bles, *Studies Biol. Lab. Owens Coll.* II.; cf. Giles, *Q. J. M. S.* XXIX.

are known to dispose of harmful bacteria that have gained entrance into the body. It is from some of these colourless corpuscles that a ferment, fibrin-ferment, is set free which converts a soluble proteid in the plasma into solid, insoluble fibrin, and causes the blood to clot on wounded surfaces, and so staunch the bleeding. The blood is contained in a system of well-defined vessels—heart, arteries, capillaries and veins; there are no extensive, irregular, vascular spaces such as occur in molluscs and arthropods. With this system the lymphatic vessels are in communication in a few places, and it is said that on the ventral surface of the kidney there is direct connexion between the cœlomic space and the cavities of the tributaries of the renal vein.

The fact that the blood is kept in motion may be easily verified by examining the web of the foot of the adult, or the tail or external gills of the tadpole, under the microscope. The heart, which is the force-pump for maintaining the flow, lies in the anterior part of the body, and is protected from external shock and pressure by the sternum and ventral portions of the shoulder-girdle. It consists of, (1) a dorsally situated sinus venosus, into which the blood from the greater part of the body is poured by three great veins, the two anterior and the single posterior vena cava; (2) the right and left auricles, forming the wider anterior part of the heart; the right is the larger, and receives the blood from the sinus venosus; into the left arterial blood is poured by the pulmonary vein; the two auricles are completely separated by a thin partition; (3) a single muscular ventricle, bluntly conical in shape, posterior to the auricles; it receives blood from both auricles; (4) the truncus arteriosus, which leads out from the right anterior ventral corner of the ventricle.

The blood passes through the several chambers in the above order, and is prevented from going in the contrary direction, when the walls of the heart contract, by valves placed at the auriculo-ventricular openings and on the truncus arteriosus, and by the fact that at each beat of the heart a wave of contraction sweeps continuously over the whole set of chambers commencing at the sinus venosus and ending at the truncus arteriosus. Thus the contraction itself urges the blood in the proper direction and opposes any reflux. Since the heart continues to beat for some time after general death the various phases and appearances can be readily observed.

The ventricle contains both venous and arterial blood, venous on the right and arterial upon the left side, while the two to a certain extent mingle in the centre. There is however no doubt that much of the blood returned from the skin and mouth into the right auricle is fairly rich in oxygen (*vide* p. 261, *Respiration*). The valves within the truncus arteriosus are so disposed that the most venous blood is directed along the pulmo-cutaneous arteries to the lungs and skin. The mixed blood is conveyed by the systemic aortæ to all the rest of the body except the head, which receives the most arterial blood by the carotid arteries. Thus the brain is supplied with better oxygenated blood than any other part of the body.

The two systemic aortæ sweep round to the dorsal side of the gullet, and unite posteriorly to form the dorsal aorta. Close to their junction is given off a large vessel which conveys blood to the digestive viscera. The dorsal aorta passes backwards, giving off arteries to the kidneys and back, and eventually bifurcates into large arteries, one for each leg. The blood which passes through the capillaries of the intestines, stomach and spleen is gathered up into the hepatic portal vein which passes to the liver, and there breaks into capillaries which penetrate between the liver cells. It is by this arrangement that the food absorbed from the digestive organs is temporarily deposited as glycogen and fat within the liver by the activity of the liver cells. There is also brought to and distributed in the liver by the anterior abdominal vein blood from the hind limbs, bladder, and ventral body wall. The remainder of the blood from the hind legs is taken by the sciatic and iliac veins into the renal portal veins, which, running up the outer border of the kidneys right and left, break into spacious capillaries within the substance of the kidneys themselves. Here nitrogenous waste matter is eliminated from the blood, which then issues into the posterior vena cava, and is thus returned to the heart.

The lymphatic portion of the circulatory system is very extensively developed. In addition to the lymphatic vessels and capillaries that permeate all the organs of the body there are between the skin and the underlying muscles numerous large lymph sacs separated from one another by thin connective tissue partitions which pass from the skin to the muscles, and which are constant in their positions. It is the presence of these sacs which makes the frog's skin appear to fit so loosely. The lymph within them is probably an important factor in keeping the skin moist and in a fit state for the interchange of gases in respiration. There are numerous minute openings between these subcutaneous lymph sacs and the adjacent lymphatics of the underlying organs. In connexion with the vessels are two pairs of small contractile sacs, the anterior and posterior lymph-hearts: the former lie behind the transverse processes of the third vertebra and by their rhythmic contractions, assisted by sundry valves, drive lymph into the vertebral vein; the latter are placed on either side of the hinder end of the urostyle—the rod of bone which continues the line of the vertebral column—and pour their contents into a vein that connects the femoral and iliac veins near the top of the thigh. The lymph itself is a clear transparent fluid containing colourless corpuscles, identical with those of the blood. It is virtually blood *minus* red corpuscles. Thus all the solid tissues of the body are constantly bathed in fluids bringing to them from the respiratory and digestive organs oxygen and nutrient material, and removing carbon dioxide and other waste products which are conveyed away to the excretory organs, and by them cast out of the body.

Respiration[1]. The respiratory mechanism is one of considerable complexity. The absence of diaphragm and of ribs at once shows that inspiration and expiration are carried on by means very different from those found in ourselves. Moreover the skin of a frog is thin and constantly moist, so that much of the interchange of gases

[1] Gaupp, *Arch. f. Anat. und Phys.* (*Anat. Abth.*), 1896; Baglioni, *ibid.* (*Physiol. Abth.*), 1900, Supp. Bd.

is effected through it without the active intervention of any muscular apparatus. Indeed during hibernation the skin alone is sufficient for all respiratory needs, thanks to the copious supply of blood brought to it by the cutaneous branch of the pulmo-cutaneous artery. During the periods of normal activity, however, the respiratory movements are very evident. These consist of rapid superficial oscillations of the throat, interrupted by occasional more profound movements affecting the ventral surface and flanks of the body. The up-and-down movements of the throat take place while the glottis (the entry to the passage leading to the lungs) is closed, and are only accompanied by slight movements of the nostrils, which are for the most part open; they have no direct connexion with the pulmonary respiration. Mascacci and Camerano have shown that an interchange of gases takes place through the mucous membrane of the mouth, and that this is quite as important a respiratory organ as the skin[1]. The latter of these two observers finds that in certain lungless amphibians, *e.g. Spelerpes fuscus* and *Salamandrina perspicibata,* buccal respiration has entirely replaced pulmonary, while the cutaneous interchange is quite unimportant; and in these forms the oscillations of the throat are particularly well marked. It is probable that the peculiar arrangement of the capillary blood vessels of the mucous membrane of the mouth and anterior part of the œsophagus is intimately connected with this method of breathing. Langer and Schöbl have described these capillaries in the frog as arranged in a

[1] Cf. Howes, *Jour. Anat. and Phys.* XXIII.

close meshwork with frequent irregular dilatations of the walls. This peculiarity would cause the blood to pass very slowly through the capillary plexus and give more time for the respiratory interchange of gases.

To understand the more extensive muscular movements by which the pulmonary respiration is brought about, it is necessary to bear in mind, (1) that the lungs are elastic and that their natural tendency is to expel air from their cavities; (2) that there are no firm thoracic walls by which pressure may be removed from the outer surface of the lungs, and therefore (3) that air must be forced down into the lungs by some mechanism acting along the air passages themselves with sufficient power to overcome the elastic recoil of the walls of the lungs. For our knowledge of the phenomena we are chiefly indebted to the researches of Bert[1], Gaupp[2] and Baglioni[2]. Three distinct phases are recognisable in these movements, (1) aspiration of air through the nostrils into the cavity of the mouth: this is brought about by lowering the floor of the mouth, the glottis remaining in its resting position, *i.e.* closed; (2) expiration of air from the lungs into the enlarged mouth cavity; this is achieved by opening the glottis, and by the contraction of the abdominal muscles and of the walls of the lungs; (3) inspiration or injection of air into the lungs by closing the nostrils, keeping the glottis open, and raising the floor of the mouth. This action can be detected externally by the

[1] *Leçons sur la Physiol. comparés de la respiration*, Paris, 1870.

[2] Gaupp, *Arch. f. Anat. und Phys.* (*Anat. Abth.*), 1896; Baglioni, *ibid.* (*Physiol. Abth.*), 1900, Supp. Bd.

sudden outward movement of the skin covering the tympanic membranes. The increased pressure of air within the oral cavity, acting up the Eustachian recesses, forces the membrane outwards so that it becomes for a moment slightly convex. The two last phases follow one another in the order stated and with extreme rapidity. It will thus be seen that the air in the lungs is never entirely renewed, but is merely refreshed by mixing with that aspired into the mouth cavity through the nostrils. The mouth itself is kept firmly closed throughout. Before the introduction of air into the lung is quite complete the nostrils are opened again, or they may be kept slightly open from the commencement of this phase; in either event the elevation of the floor of the mouth expels a certain amount of air from the body through the nostrils. By keeping the nostrils completely closed during "injection" the animal can inflate itself to some extent. On the other hand if the nostrils are artificially or naturally kept open during the complete act of respiration the amount of air in the body is diminished.

The muscles by which the floor of the mouth is lowered and pulled backwards in aspiration are the *sternohyoids* arising from the coracoid and clavicle and inserted on to the ventral surface of the body of the hyoid, and the *omohyoids* arising from the anterior border of the scapula and inserted on to the outer part of the ventral surface of the body of the hyoid. The contraction of these muscles pulls the flat plate of cartilage, known as the body of the hyoid, which forms the floor of the mouth downwards and backwards. Thus the cavity of the mouth is enlarged and air rushes in through the nostrils. At the same time by the contraction of the same muscles the hyoid apparatus is drawn against the lungs so as to compress them. Hence the air in the lungs begins to pass through the glottis directly the latter is opened.

Expiration is in part due to the muscles just mentioned, but is caused chiefly by the contraction of the abdominal muscles, particularly the anterior portions of those which run transversely, and by the elasticity of the lungs themselves.

Injection of the air into the lungs is accomplished primarily by the *petrohyoid muscles*: there are four of these on each side of the head; the most anterior is a fairly wide sheet, the other three mere slips; they rise from the outer surface of the auditory capsule, pass fan-wise round the floor of the pharynx and œsophagus and are inserted into the mid ventral line of the pharynx and the sides of the hyoid. The effect of their contraction is therefore to draw the floor of the mouth and pharynx upwards and forwards, and thus forcibly to compress the air contained in the mouth, driving it through the open glottis down into the lungs. At the same time the muscles of the lower jaw contract vigorously, and assistance may also be given by the *mylohyoid, geniohyoid* and other muscles in the floor of the mouth itself.

The lateral margins of the nostrils are bounded by the shell-shaped alar cartilages which are moveably attached by their anterior ends to the nasal processes of the premaxillary bones.

The closure of the nostrils appears to be brought about by the firm closure of the jaws. The pressure of the lower jaw bends up the intermaxillary region of the upper jaw, forcing the alar cartilages against the inner margin of the nostril, temporarily rendered tense by certain muscles[1], and thus shuts the nostrils externally. Their internal apertures are covered by the two processes *A*, *A* (Fig. 44) which arise from the anterior sides of the bases of the anterior cornua of the hyoid, and which are pressed up against them.

The glottis is normally closed by its own elasticity; the muscles which effect its opening originate from the bony posterior cornua of the hyoid and are inserted along the lips of the glottis at right angles to the axis of the slit.

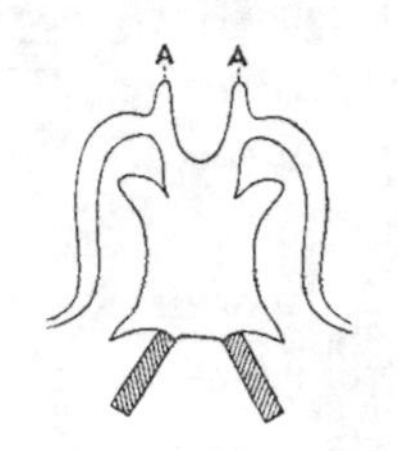

Fig. 44. The Hyoid apparatus of a frog. The posterior cornua are shaded.

At the end of each act of pulmonary respiration the rapid oscillatory movements of the throat are resumed with the nostrils open. In this way the mixture of fresh and expired air left in the mouth and throat is quickly exchanged for air entirely fresh.

The several phases and actions in pulmonary respiration may be summarised as follows:

1. Aspiration. Nostrils open; glottis closed; floor of mouth lowered by *sterno-* and *omohyoids.*

2. Expiration. Nostrils closed; glottis open; lungs compressed by action of *sterno-* and *omohyoids,* by abdominal muscles, and by intrinsic elastic tissue.

[1] Bruner, *Anat. Anzeig.* XV.

3. Inspiration (Injection). Nostrils closed or open; glottis open; air forced into lungs by *petrohyoids* and other muscles, while some escapes through nostrils.

The Nervous System and Axial Skeleton. The nervous system of the frog is far more complicated, and the functions of its component parts more fully understood than in the case of any of the preceding animals described in this volume. It consists of a well-developed brain and spinal cord from both of which nerves proceed to the sense organs, muscles, and other viscera. The brain is enclosed within the cranium, which is an almost complete tube of cartilage. In parts it is ossified, elsewhere plates of bone, developed in membrane, are laid down outside the cartilage. The brain is thus adequately protected from external shock, and a firm axis is afforded upon which the bones of the jaws are fixed, and on which those which are moveable find fulcra more or less directly in their movements as levers: it also serves as a fixed base upon which the muscles of the eyes and jaws and other anterior parts of the body can pull. In the same way the spinal cord, which is continuous with the hinder part of the brain, is protected by the vertebræ of the "backbone." During early life the spinal cord is supported by the notochord; subsequently this flexible rod-like structure is supplanted by the rigid bony centra of the nine vertebræ and the unsegmented urostyle. Flexibility of the trunk is however retained by the free articulations of the vertebræ one with another. From the centra there spring corresponding arches of bone—the neural arches—which enclose the spinal cord and roof it over as with a

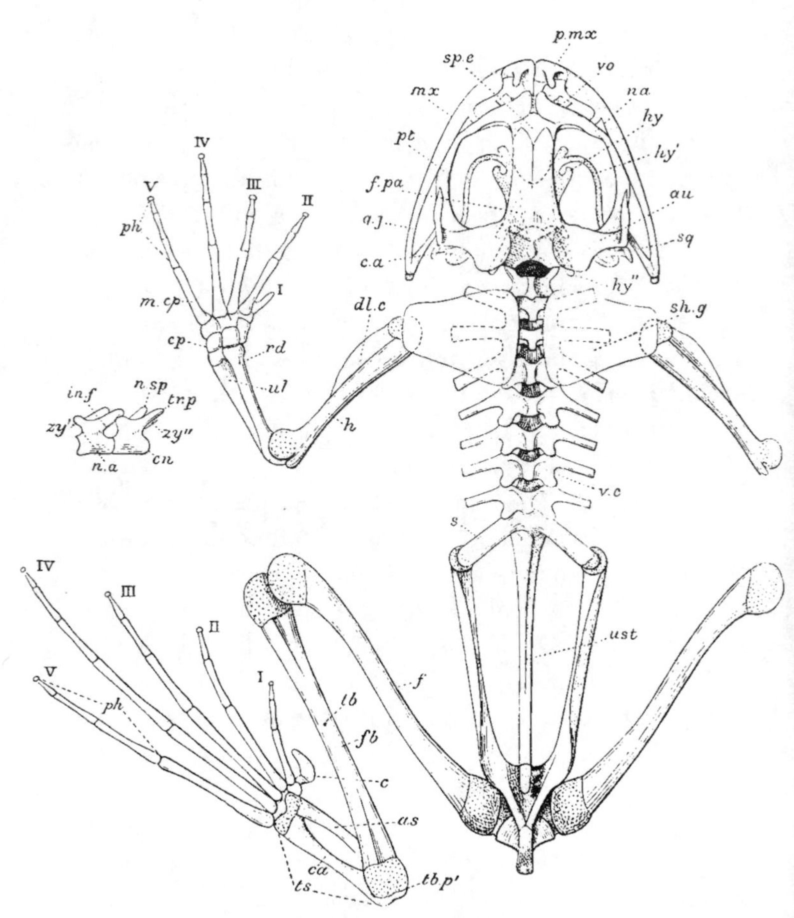

Fig. 45. The entire axial skeleton, with the limb-girdles and the limbs of the left side, drawn from above with the body in the resting attitude. To the left are shown the fifth and sixth vertebræ, drawn from the left side to show their natural relation. After Howes.

au. Periotic capsule. *as.* Tibiale (Astragalus). *c.* Calcar. *c.a.* Columella auris. *ca.* Fibulare (Calcaneum). *cn.* Vertebral centrum. *cp.* Carpus. *dl.c.* Deltoid crest. *f.* Femur. *fb.* Fibula. *f.pa.* Fronto-parietal. *h.* Humerus. *hy.* Body of the hyoid. *hy'*. Its anterior cornu. *hy''*. Its posterior cornu (thyro-hyoid). *in.f.* Intervertebral foramen. *m.cp.* Metacarpus. *mx.* Maxilla. *na.* Nasal. *n.a.* Neural arch. *n.sp.* Neural spine. *ph.* Phalanges. *p.mx.* Pre-maxilla. *pt.* Pterygoid bone. *q.j.* Quadrato-jugal. *rd.* Radius. *s.* Sacrum. *sh.g.* Shoulder-girdle. *sp.e.* Sphenethmoid. *sq.* Squamosal. *tb.* Tibia. *tb.p'.* Tibio-tarsal joint. *tr.p.* Transverse process. *ts.* Tarsus. *ul.* Ulna. *ust.* Urostyle. *v.c.* Vertebral column. *vo.* Vomer. *zy'.* Anterior zygapophysis. *zy''.* Posterior zygapophysis. I to V. Digits i to v.

tunnel. Bony outgrowths for the attachment of muscles arise from the neural arches: each vertebra bears upon the summit of the arch a low crest, the neural process, and, with the exception of the first or atlas vertebra, a pair of transverse processes, one on the right and the other on the left side, at the junction of the arch with the centrum. To these processes and also to the sides of the urostyle numerous muscles are attached, while to the stout transverse processes of the ninth vertebra are firmly fastened the long ilia of the pelvic girdle. The pelvic girdle connects the hind leg with the trunk, and it is thus through the ilia that the movements of the legs are transferred to the body as a whole, hence in leaping and swimming severe strains are thrown upon the ninth pair of transverse processes, and extra strength is required in them.

The head is capable of movement up and down upon the atlas vertebra, the two smooth occipital condyles working in corresponding concavities in the anterior face of the centrum of the atlas. A limited amount of "play" is permitted between the successive vertebræ by their articulations one with another. The centra fit into one another by ball- and socket-joints: the first seven vertebræ have the ball at the hinder end, and the socket at the front end of the centrum, the eighth has a socket at each end, while the ninth has a ball at the front and a split knob at the back articulating with the urostyle. Further connexion between the vertebræ is afforded by interlocking processes projecting forwards and backwards from the dorsal region of the neural arches. Each vertebra, except the atlas, has two pairs of such processes, one pair anterior, the other posterior. The anterior pair of any one vertebra is overlapped by the posterior pair of the vertebra next in front. The surfaces of contact are smooth, and in life are covered by smooth cartilage; they are bound together by ligaments and, as in all moveable joints, a fluid, termed synovial fluid, acts as a lubricant to prevent friction. It will thus be seen that at the anterior end of a vertebra the smooth facets of the interlocking processes are directed dorsalwards, but at the posterior end ventralwards, and so the head- and tail-end of any isolated vertebra can be readily determined. The axial skeleton then performs two main functions, it is the main pillar in the animal fabric, and it protects the central nervous system on which depends the co-ordination of the entire machine.

The brain itself consists, beginning at the anterior end, of the olfactory lobes, of the cerebral hemispheres, the thalamencephalon, with the pineal body, infundibulum and pituitary body, the optic lobes, the cerebellum and the medulla oblongata (*vide* Fig. 46). The cerebral hemispheres are the seat of origin of all spontaneous actions, that is to say, actions not provoked reflexly by external circumstances. If the hemispheres are removed the frog still swims when put in water, crawls so as to rest on a horizontal surface, avoids obstacles in its path, swallows and breathes. But none of these actions are performed except the appropriate stimulus be applied, and on the other hand any given stimulus is followed with mechanical certainty by the appropriate action. The hemispheres then are the seat of the will; the other portions of the brain are concerned with the receipt of impulses from without, and with the co-ordination of muscular actions. The spinal cord of itself, after severance from the brain, is able to carry out many complex reflex actions and these often of a purposeful character: in life it is also the main road of communication between the brain and the spinal nerves which are distributed throughout the body. From the brain there are given off ten pairs of cranial nerves which pass through the walls of the cranium to their several destinations. The subjoined table (p. 273) shows their names and functions.

The spinal cord gives off also ten pairs of nerves which pass out between the successive vertebræ. Each arises by two roots, a dorsal and a ventral, which unite to form one nerve. The dorsal root bears upon it a ganglionic enlargement, and is composed of nerve fibres conveying sensory impulses only; the ventral root on the other hand contains only motor fibres. Conspicuous upon the spinal ganglia are white calcareous patches—the periganglionic glands or calcareous sacs. Lenhossék[1] states that there are usually two to each ganglion, covering its sides and

[1] *Arch. f. mikrosk. Anat.* XXVI.

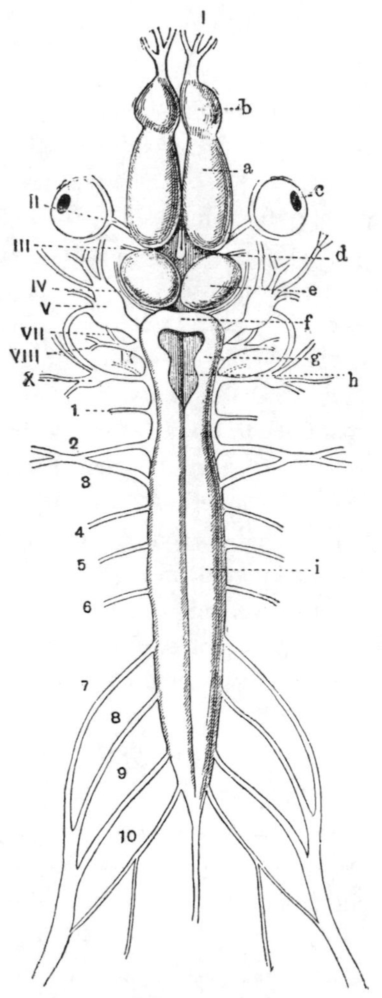

Fig. 46. Brain and spinal cord of frog × about 2.

a. Cerebral hemisphere. b. Olfactory lobe. c. Eye. d. Thalamencephalon. e. Optic lobes. f. Cerebellum. g. Medulla oblongata. h. Fourth ventricle. i. Spinal cord. I. Olfactory nerves. II. Optic nerve. III. Oculomotor nerve. IV. Patheticus. V. Fifth nerve. VII. Facial nerve. VIII. Auditory nerve. IX. Glossopharyngeal nerve. X. Vagus nerve. 1—10. First to tenth spinal nerves. 2 and 3 unite to form the brachial, and 7, 8, and 9, to form the sciatic plexus. [The sixth cranial nerve is very small and is not shown.]

Name of Nerve	Destination	Function
1. The olfactory	Mucous membrane of nostril	Conveys to brain impulses causing sensation of smell (sensory)
2. The optic	Retina of eye	Conveys to brain impulses causing sensation of sight
3. The oculo-motor	Rectus superior, ,, inferior, ,, internus, Obliquus inferior } muscles of the eyeball	Conveys from brain impulses causing these muscles to contract and move the eyeball (motor)
4. The pathetic	Obliquus superior ,,	,, ,, ,,
5. The trigeminal	(a) Mucous membrane of nose and skin of front part of head (b) Wall of eyeball, mucous membrane of mouth (c) Upper eyelid and surrounding skin; temporal and pterygoid muscles; skin of lower eyelid, upper lip and of side of head; skin of lower lip and depressor maxillæ; mylohyoid and submental muscles, under surface of floor of mouth and lower lip	Sensory for skin and mucous membrane. Motor for muscles supplied
6. The abducens	Rectus externus and retractor bulbi } muscles of the eyeball	Motor
7. The facial	Mucous membrane of roof of mouth; muscles of anterior cornu of hyoid and skin of throat and sternal region; skin of tympanic membrane and behind angle of mouth	Sensory and motor
8. The auditory	The internal ear	Sensory (nerve of hearing)
9. The glossopharyngeal	Petrohyoid muscles, and mucous membrane of pharynx and tongue	Sensory and motor
10. The pneumogastric or vagus	Muscles of the back, the larynx, the heart, the lungs, the stomach	Motor to muscles supplied. Regulates heart-beat and respiration

ventral surface. The glands consist of numerous tubes whose epithelium secretes a milky calcareous fluid: they are highly vascular and appear in the tadpole prior to the development of the limbs. Their function is unknown.

A spinal nerve then, after the union of the roots, is a mixed motor and sensory nerve. These ten pairs of nerves are distributed to nearly all parts of the body not supplied by the cranial nerves. The remainder is innervated by what is known as the Sympathetic System. This consists of a row of nerve ganglia, connected by nerve fibres and lying on either side of the vertebral column but continued forward inside the cranial cavity. One sympathetic ganglion is usually associated with each spinal nerve and united to it by a "ramus communicans"; with the tenth spinal nerve more than one ganglion may be connected. The right and left cords are united to one another by numerous fine nerve strands which surround the dorsal aorta and supply fibres to the neighbouring vessels and organs. Sympathetic fibres run with the spinal nerves to all parts of the body, and there are special bundles of them passing to the heart, stomach, liver, and other abdominal viscera. They are especially concerned with regulating the calibre of blood vessels and the degree of contraction of the muscular walls of hollow viscera in general. The heart possesses the power of continuing to beat after all its nerve connexions have been severed, and indeed after it has been removed from the body: its rate of beating is however under nervous control by means of the various nerve fibres reaching it by the vagus and sympathetic branches.

Excretory System. The excretion of the nitrogenous waste substances produced throughout the body is performed by the two kidneys. These are reddish-brown organs of an elongate-semilunar shape, lying parallel to the vertebræ close to the dorsal wall of the abdomen, and outside the peritoneal lining. They receive blood both from the renal arteries and from the renal portal veins. The renal veins conduct the blood from them into the posterior vena cava. The kidney substance itself is composed of very numerous uriniferous tubes, which excrete water and urea from the blood. Each tube commences blindly within the ventral portion of the kidney in a globular dilatation—the Malpighian capsule. The cavity of the capsule is largely occupied by a cluster of blood vessels—the glomerulus—derived from a twig in the renal artery. Strictly speaking the glomerulus is external to the capsule; it appears to be inside it in consequence of one-half of the capsule being thrust inward inside the other, in the same way as one-half of a soft rubber ball may be tucked inside the other by pressure with the finger. The course of the individual tubes, though very tortuous, is constant. The epithelium of the first portion of the tube is ciliated, then follows a glandular portion whose cells contain a yellow pigment and are said to bear short cilia (Tornier[1]): the next part is densely ciliated and passes into a wider portion, the protoplasm of whose cells shows a rod-like structure. This part opens at right angles into a collecting tube. The collecting tubes run across the dorsal surface of the kidney into the ureter, which,

[1] *Arch. f. mikrosk. Anat.* XXVII.

beginning anteriorly on the dorsal surface, passes to the outer edge of the kidney and runs backward close to its fellow into the dorsal wall of the cloaca.

The epithelium of the capsule removes water from the blood; nitrogenous compounds are eliminated by the glandular portions of the tubes, and become dissolved in the water which is swept along the tube by the action of the cilia.

Some authorities have described peritoneal funnels (nephrostomes) to the number of about 300 upon the ventral surface of the kidneys. According to Spengel[1] and Meyer[2] these open upon the ventral surface and are connected with the last portions of the uriniferous tubes: Nussbaum[3] however states that they open into the first ciliated part in the young larvæ. On the other hand, Wiedersheim maintains that the funnels, though present, do *not* open at the surface into the abdominal cavity, while Heidenhain[4] and Ecker[5] were entirely unable to find any such structures. There is no doubt that nephrostomes exist in the tadpole and that they lead into the uriniferous tubes. In the adult, according to Nussbaum, Milnes Marshall[6], and others, whose experiments appear to be conclusive, they establish connexion with the intra-renal blood vessels. It is clear that the condition of these funnels varies with the age of the individual.

[1] *Centralbl. f. d. med. Wiss.* 1875, and *Arbeit. a. d. Zool. Zoot., Instit. der Univ. Würzburg*, Bd. III. 1876.

[2] *Sitz. d. nat. Gesell.*, Leipzig, 1874.

[3] *Sitz. d. Niederrhein. Gesell. in Bonn*, 1877, XXXIV.

[4] *Arch. f. mikrosk. Anat.* 1874, X.

[5] *The Anatomy of the Frog.*

[6] *Stud. Biol.: Lab. Owens Coll.* II.

The ureter of the male has upon it an enlargement, the *vesicula seminalis* (*vide infra, sub Reproduction*).

The urinary bladder is not directly connected with the ureters. It is a bilobed, thin-walled sac lying ventral to the cloaca and opening through a narrow neck into the cloacal portion of the alimentary canal. It has been asserted[1] that the liquid contained in the bladder is not urine but water only, and that the organ is for the storage of water for the respiratory functions of the skin There is however no doubt that urea is present in the liquid. When alarmed the animal usually voids the contents of the bladder and may perhaps gain some protection by this means of repelling attack.

Attached to the renal veins on the ventral surface of the kidney are small yellow bodies, the adrenals. Srdinko[2] regards these as blood-forming and cytogenous organs.

Reproduction. The sexes are separate, though abnormal instances of hermaphroditism are not unknown in which the reproductive organs partake of the characters both of testis and ovary. The testes lie upon the ventral surface of the kidneys; they are of a creamy-white colour with black pigment spots freely scattered over their surface. In shape they are ovo-spherical as a rule. Their size varies considerably at different seasons, being greatest at the breeding time. The *vasa efferentia*, or sperm ducts, pass into the kidney, where they open into a longitudinal tube, whence transverse tubes arise which unite with uriniferous

[1] Bell, *British Reptiles*.

[2] *Anat. Anz.* XVIII.; cf. Swale Vincent, *ibid.*

tubes, and so lead into the ureter. In the male sex then the ureter functions also as *vas deferens*, conveying not only the urine but also the spermatozoa. In this sex, as

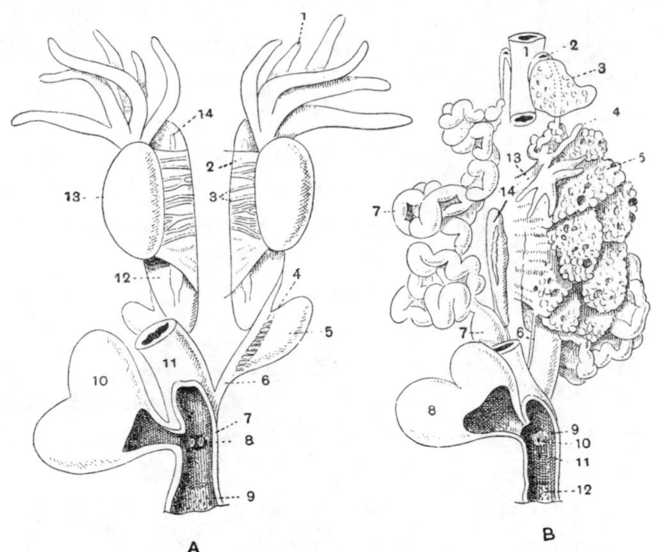

Fig. 47.

A. The urino-genital organs of the male, dissected from the front, after removal from the body. From Howes.

B. The urino-genital organs of the female, dealt with in the same manner as the above, except that, in order to show the natural relations of the mouth of the oviduct, the left lung and a portion of the œsophagus were also removed from the body.

A. 1. Fat-body. 2. Fold of peritoneum supporting the testis. 3. Efferent ducts of testis. 4. Ducts of vesicula seminalis. 5. Vesicula seminalis. 6. Ureter. 7. Cloaca. 8. Orifice of ureter. 9. Proctodæum. 10. Allantoic bladder. 11. Rectum. 12. Kidney. 13. Testis. 14. Adrenal body.

B. 1. Œsophagus. 2. Mouth of oviduct. 3. Left lung. 4. Corpus adiposum. 5. Left ovary. 6. Ureter. 7. Oviduct. 8. Allantoic bladder. 9. Cloaca. 10. Aperture of oviduct. 11. Orifice of Ureter. 12. Proctodæum. 13. Fold of peritoneum supporting the ovary. 14. Kidney.

already mentioned, there is upon the outer side of each ureter and connected with it by numerous fine ducts a considerable dilatation, the *vesicula seminalis*, within which spermatozoa can be stored; the walls of the *vesicula* also secrete a fluid in which the spermatozoa swim. Closely united with the external edge of the *vesicula* is a fine cord which passes forwards, tapering away, and represents the oviduct of the female.

The ovaries in the female occupy positions corresponding to the testes of the male. They are however very much larger and are divided into about ten well marked lobes. At the breeding season they become immense and occupy the greater part of the body cavity. They present, when nearing maturity, a speckled black and white appearance owing to the ova being black at one pole and white at the other. The oviducts are, in the full-grown animal, highly convoluted, white, opalescent tubes. They open by semilunar slits into the body cavity at its anterior end near the bases of the lungs. The two apertures face obliquely towards one another and are lined with ciliated epithelium. The walls of the oviducts are glandular, and the lumen of the tube of almost uniform calibre until just before they reach the cloaca. At this point they suddenly enlarge and the walls become thin. The oviducts open into the dorsal wall of the cloaca, the right slightly behind the left, immediately anterior to the openings of the ureters.

The breeding season may begin as early as February or, if the winter be prolonged, as late as April. The male is now easily distinguishable from the female by his smaller size and less cumbrous figure, the body of the female being greatly distended by the ripe ova and enlarged oviducts. Moreover the coloration now differs in the two sexes. The dorsal surface of the male becomes a uniform dark olive-brown colour, and the temporal blotch almost ceases to be distinguishable: the ventral surface, and especially the throat, becomes extraordinarily pale and is a faintly bluish grey colour. On the other hand, the pigments of the female are sharply contrasted; the general colour is a yellowish brown relieved by small spots of bright yellow: the temporal blotch is more than usually conspicuous and the ventral side of the body, particularly the throat, is coloured as warmly as the flanks and dorsal surface. In both sexes a cresentic mark of golden yellow appears upon the skin bordering the nasal margin of the eye-ball. The innermost (true 2nd) digit also of the male is at this season modified for mating purposes. This digit is normally somewhat swollen at the base but during the breeding season the swelling enlarges and becomes of a black-brown colour; the epidermis covering it becomes thickened and papillous. The enlargement is chiefly caused by the increase in size of the mucous glands of the skin in this region. The papillæ are peculiarly rich in "touch-spots" —nerve structures which also occur on the dorsal surface of the trunk and under-surface of the hind feet.

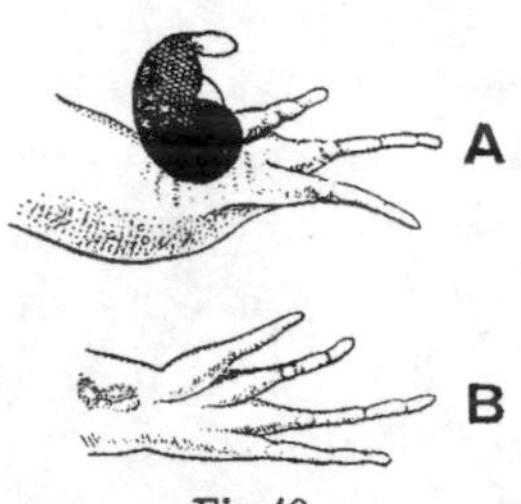

Fig 48.

A. The left manus of *Rana temporaria*. ♂ Palmar surface.

B. The same of the ♀.

As soon as the weather has become sufficiently warm the Frogs, which have hibernated some in holes and drain-pipes, others in or on the mud at the bottom of ponds, assemble in suitable bits of water and proceed to pair. In this act the male mounts upon the back of the female clasping her firmly round the pectoral girdle, just behind the arm-pits, with his fore limbs. It is said that to render the grasp more secure the modified digits, which stand, after the fashion of a thumb, at an angle to the others, are hooked over one another and further united by the mucus of the enlarged glands. This statement I cannot confirm. I find that the two digits in question are pressed firmly towards each other, nipping a piece of the skin of the female between them. The protuberances upon them are pressed strongly upwards against the body of the female on either side of the sternum. It is possible that their pressure may in some way assist in the passage of ova into the oviducts. In this position the animals remain for many days prior to oviposition. Eventually, in the sunshine of early morning, the female discharges the eggs into the water while the male pours spermatozoa over them as they issue.

Correlated with this use of the male index digit are certain more deeply seated modifications which affect the skeleton. The male second metacarpal bone is stronger and broader than that of the female, and has upon its inner side a projection for the insertion of the muscle (*abductor digiti* II *longus*) which is employed in maintaining the two front feet pressed together. Similarly the lower half of the humerus of the male bears upon its posterior surface a ridge which reaches down to the inner condyle and is especially prominent at the breeding season, whereas that of the female is cylindrical. From this ridge arises a muscle (*flexor carpi radialis*) which is also in vigorous action while the male embraces the body of his mate. While pairing is in progress and prior to it the male frogs often keep up a chorus of croaking. At a distance of twenty or thirty yards the noise made by a number all croaking together is not displeasing, and might be mistaken for the cooing of a pigeon. The females also croak loudly and appear to inflate the floor of the mouth as they do so. It is noteworthy that the animals can croak when several inches below the surface of the water, and that the noise can even then be heard at a distance of several yards. The noise is produced by the vibration of two vertical flat bands of connective tissue, the vocal cords, which project into the cavity of the larynx, or upper portion of the very short windpipe; the position and tension of these cords can be varied and regulated by certain muscles which move the cartilages to which the cords are attached.

When the ovaries are ripe they shed their contents into the body cavity. The exact method by which the ova find their way into the oviducts is not fully understood. According to Nussbaum[1] in *Rana fusca* certain parts of the peritoneal lining of the body cavity are ciliated. Presumably by the action of the cilia and possibly also by contraction of the abdominal muscles the ova are driven forward into the region of the oviducal apertures and eventually forced into them. The ciliated internal lining of the oviducts forces the ova backwards along the convoluted tubes towards the cloaca. In course of transit

[1] *Arch. f. mikrosk. Anat.* XLVI.

the oviducal glands deposit a layer of mucus around each ovum. On reaching the water this substance swells enormously by absorption of water and forms the glairy, transparent, slippery covering which surrounds each egg in the familiar masses of frog-spawn[1].

The functions of this mucous envelope are manifold; in process of swelling it appears to exert a powerful attractive influence upon the spermatozoa and thus aids in fertilisation; it resists putrefactive changes for a very long while and in this way shields the developing tadpole from bacterial attack; it holds the eggs together, rendering the mass difficult to transport in the water by increasing the probability of entanglement in weeds should there be any current flowing along the ditch or through the pond in which spawning has occurred; it spaces the eggs from each other, allowing sufficient oxygen to be available to each, and entangles in its interspaces masses of green swarm-spores of Algae which give out oxygen when exposed to sunlight and thus help to aërate the eggs; it interposes an impenetrable barrier between the eggs and the small carnivorous Crustacea which abound in fresh water, and by its slipperiness protects the eggs from the attacks of aquatic birds. I have however seen newts, when very hungry, attack masses of frogs' spawn and gulp down portions of it with considerable difficulty: nevertheless they succeeded in so demolishing the mass that none of the eggs produced tadpoles. I am inclined to think that there may be some unpleasant flavour about

[1] If held up to the light the outer layer of mucus is seen to be more transparent and less dense than the inner.

the mucus, for I have seen toads at the spawning season killed in large numbers by crows who attacked them before spawning had actually been completed: the birds picked the toads to pieces, devouring all the muscles, liver and intestines but in every case they left the skin, which is known to be extremely acrid, the oviducts and such ova as were coated with mucus—of the ovaries themselves not a trace could be found. Some of the spawn had already issued in its characteristic ropes, and was still continuous with other portions lying within the oviducts.

In its normal condition frogs' spawn floats in the water. It is probably buoyed up to some extent by bubbles of oxygen arising from submerged water plants. If removed from the water and carried in a small vessel it will frequently sink when placed in an aquarium: this however does not seriously interfere with the normal development of the young. The natural position at the surface of the water is however of importance. It places the eggs in the warmth of the sun, and it is noteworthy that in ditches the spawn is usually deposited against the more sunny bank. It is possible that to a small extent the convex surface of the mucous envelopes concentrates the rays of the sun upon the eggs within. Each egg is black in its upper half and, at first, white beneath; hence the surface exposed to the sun has the greater power of absorbing heat.

The precise rate of development of the fertilised egg depends to some extent on the temperature, so that the periods mentioned in the following paragraphs must be taken as averages. In about four days the black colour has spread over the whole egg. By the end of the first week after fertilisation the shape is no longer spherical but decidedly ovoid, the embryo becoming much longer than broad. By the tenth and eleventh day head, body, and tail are recognisable. The tail grows rapidly, and from the sides of the throat, two and then a third pair of external gills project as branching filaments. The entire growth thus far takes place at the expense of the food-yolk contained within the egg at the time of deposition. About a fortnight after fertilisation of the eggs the young tadpoles escape from the gelatinous mucous envelopes and swim in the water by means of their well-developed tail. They are now from a third to half an inch long. The whole surface of the body is

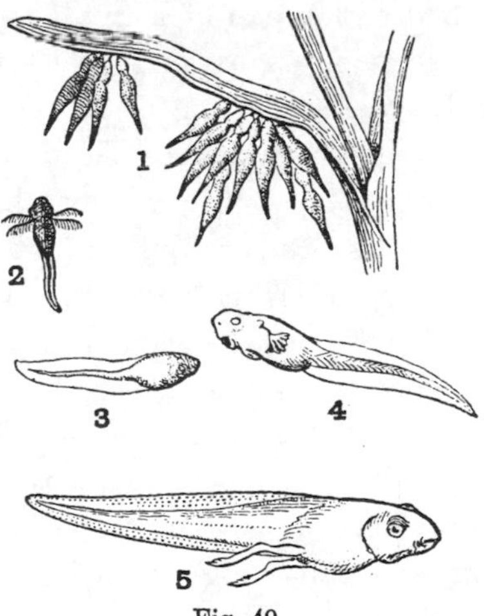

Fig. 49.

1. Tadpoles soon after hatching, clinging to water-weeds. 2. Tadpole with two pairs of external gills. 3 and 4. Tadpoles with operculum forward and forming. 5. Tadpole with well-developed hind legs.

covered with cilia, which keep the water round the tadpole in constant motion, and thereby assist respiration. Upon the ventral surface of the head is a horseshoe shaped structure commonly called the sucker, but in reality consisting of a mass of mucous glands by whose secretion the animal attaches itself to weeds and other objects. If a clean needle or fine glass rod be presented to a tadpole removed from the water the creature will readily adhere to it and subsequently a deposit of mucus can be seen upon the surface by the aid of a microscope. There is no mouth present until the tadpole has been hatched a few days. Notwithstanding this fact there is a decided increase in size; possibly food may be absorbed by the "suckers." When the perforation is effected the mouth is furnished with a pair of horny jaws and fringing lips. The tadpole now begins for the first time to take food for itself, its diet consisting of the leaves of water weeds and other vegetable matter. The alimentary canal grows rapidly and attains a great length so that it falls into several close coils within the abdomen. The direction of the coiling is not continuously the same; about one and a half turns take place "counter-clockwise," and about the same number "clockwise," the spiral being reversed near the middle of its course. The reversed spiral is a mechanical necessity when a tube whose ends are not free to revolve is thrown into coils. Eventually the canal becomes actually longer than it is in the young frog.

If examined under the microscope at this stage the gills are seen to be nearly transparent tubes with a solid axis; the blood may be seen coursing up one side and down the other in each filament, travelling in a jerky manner at each beat of the heart. Very soon four slits appear on each side of the throat, opening up communication between the pharynx and the outer water. The sides of these gill-slits become folded and give rise to the internal gills, and at the same time the external gills begin to decrease. Meanwhile a fold of skin, the operculum, appears in front of the gills and grows back over them so as to enclose them in a chamber. The right opercular fold grows more rapidly than the left, so that for a few days the external gills of the left side alone are visible, those of the right having disappeared beneath the operculum. At the end of about a month after hatching the opercular folds, which have united with each other ventrally, fuse also at their hinder edges with the body wall, except upon the left side where a spout-like opening is left. In its circulatory and respiratory arrangements the tadpole is essentially a fish. The blood is sent from the heart to the gills to be aërated and thence is carried all over the body before returning to the heart again. The water of respiration is taken in at the mouth, passed through the pharynx and gill-slits, where it gives up its dissolved oxygen and receives carbonic acid, and is ultimately discharged through the spout upon the left side. Further piscine characters are to be found in the cartilaginous brain-case, the muscular tail with its dorsal and ventral fins and "lateral line" sense organs, and the rod-like, unconstricted notochord which forms the axial skeleton.

In this stage the tadpole feeds voraciously and attains a length of over an inch, and now that its powers of swimming are considerable its adhesive mucous organ is of less importance and dwindles away.

At about the sixth or seventh week the hind limbs begin to appear as a pair of small warts at the base of the tail; these steadily lengthen, become jointed and, at about the eighth or ninth week, develop toes. The front limbs are not visible till somewhat later. They arise at about the same time, but are concealed beneath the operculum. They eventually become free from the covering, the right leg forcing its way through, and the left by pushing out at the spout.

In about the eleventh or twelfth week a profound change takes place in the entire economy of the tadpole. It ceases feeding; the outer layer of the skin is cast off; the eyes come to the surface; the horny jaws are lost and the fringing lips shrivel; the mouth and tongue increase in size; the gills are gradually absorbed and their function taken on by the lungs, the gill-slits closing up; corresponding changes also take place in the blood vessels and the pulmonary circulation is established. For a time both lungs and gills are functional concurrently; tadpoles that still possess gills may often be seen to rise to the surface and take air directly, rolling over so as to bring the ventral surface of the body uppermost as they do so. The character of the food is also changed, henceforth the animal is entirely carnivorous. The digestive organs undergo alteration related to this change of habit; the stomach enlarges to receive the more bulky food but the intestine becomes both shorter and narrower. When these changes are completed the tail is gradually absorbed while the hind legs lengthen so as to compensate for the loss of swimming power by reduction of the tail. The metamorphosis is then complete and the animal comes ashore, a Frog.

It is possible to defer the metamorphosis by keeping tadpoles in deep water in a vessel with vertical sides and denying all opportunities of resting the feet upon any object by which the head could be raised above the water. I have known a tadpole attain the age of two years under these conditions and even then possess but very rudimentary legs.

Enemies and Parasites. The enormous number of eggs contained in the spawn deposited by one female makes it clear that the death-rate among frogs is very high. When adult they form food for crows, herons, storks, several species of duck and occasionally some other birds; rats, moles, shrews, are known to attack them; they are the favourite diet of the grass snake and are eagerly seized by pike and other rapacious fish. No doubt when young they fall victims to all these creatures. As tadpoles they are exposed to other foes; fish so small as sticklebacks and minnows bite and kill them, but I have never seen them actually swallow any portions; newts, at any rate when pressed by hunger, will eat them, and various aquatic insects such as the larva of the Great Water-beetle, *Dyticus marginalis*, and the "Water-Boatman," *Notonecta glauca*, fasten on to them and suck them to death. Judging however by the vast number of young frogs that emerge from the water after the metamorphosis it is not during the tadpole stages that the mortality is heaviest. Internal parasites are fairly numerous. It is from the rectum of the frog that the protozoa *Nyctotherus* and *Opalina* are obtained. Both swim in the fluid contents of the rectum, and live upon the residues of the frog's food; *Nyctotherus* has a mouth and presumably has not been parasitic so long as *Opalina*, which has no aperture into its body but absorbs liquid food through its surface. This genus is multinucleate, and in reproduction the animal divides up by repeated acts of fission into as many individuals as there were nuclei at the first. The minute forms which thus result pass out at the anus of the frog into the water and obtain entry with the food into the next generation of tadpoles.

In the urinary bladder of the frog is frequently found a Trematode worm, *Polystomum integerrimum*, easily recognised by its crescent of six suckers round the posterior end. The parent, after fertilisation, protrudes her body through the urinary aperture and lays about 1000 eggs at the rate of 100 *per diem*. In some six weeks a ciliated larva, about 0·3 mm. in length, emerges and seeks, swimming by means of its cilia, a tadpole. Failing to find this the larva dies in 24 hours. If the quest be successful the young *Polystomum* enters the gill-chamber of the tadpole *viâ* the opercular spout upon the left side, and thereupon the cilia disappear. Some eight or ten weeks are spent in the branchial chamber and during this time the suckers of the adult worm are developed. At the metamorphosis of the frog the parasite makes its way into the pharynx, and thence passes along the alimentary canal to the urinary bladder, becoming sexually mature when three years old. If, however, the creature gets attached to some highly vascular, and therefore nutritive, tissue of the frog, maturity may be attained in five weeks. In such case self-fertilisation is effected and the parent dies immediately after the discharge of the eggs.

Another Trematode that may be met with in the frog is *Amphistoma subclavatum*. The early stages of this worm are passed in the water-snail *Planorbis contortus*. On escaping from the first host the miniature form encysts on the skin of a frog. Now frogs from time to time shed their skin and swallow the cast-off pellicle, stuffing it into their mouths by means of the front feet. Thanks to this habit on the part of the host the encysted *Amphistoma* reaches the stomach of the frog and there becomes sexually mature.

In the lungs and in the rectum there occurs commonly *Rhabdonema nigrovenosum*, a Nematode or Thread-worm. This parasite is free-living during part of its existence, but becomes sexually mature in the excrement of the frog. The sexes pair and the fertilised ova produce embryos which hatch within the body of the parent and devour her prior to their escape. They then live in water or may burrow into the tissues of small snails, but in neither case do they become sexually mature unless swallowed by a frog. They then make their way to the lungs and obtain a length of as much as an inch. The generation found in the lungs is hermaphrodite (successively), functioning first as male and subsequently as female. The eggs produced give rise to embryos which escape from the alimentary canal of the frog in the fæces and become the free-living form.

OTHER BRITISH AMPHIBIA.

Anura (Tailless Amphibians).

(1) The Edible Frog, *Rana esculenta*, has occasionally been found in England. It is not really indigenous but probably in all cases has been introduced. Some sixty years ago large numbers of this species and quantities of spawn were imported from the Continent and put down in the fen district. At one time they were abundant in Fowlmere Fen, Cambs. This species is larger than *R. temporaria* and greener in general colour; it has not a black mark from the eye to the shoulder, while down the middle of the back are three lines, one median and two lateral, by which it may readily be distinguished. The eyes are closer together, the intervening space more convex, the tympanic membrane larger, and the hind limbs relatively longer than in our native species. The male possesses a pair of large pouches, the vocal sacs, communicating with the cavity of the mouth and placed one on each side of the head ventral to the tympanic membrane; they are about the size of a large pea seed. The edible frog is more strictly aquatic than the common species and does not wander so far from the water.

(2) The Common Toad, *Bufo vulgaris*, is more brown and duller in colour than the frog. The skin is covered

with numerous warts and contains calcareous concretions in its dermic portions: it is thick and dry, and not at all slimy unless the animal is irritated, when an acrid mucus is discharged. This substance is of protective value, causing intense irritation to the mucous membrane of the mouth of any assailant. Dogs not infrequently suffer intense distress from taking hold of toads and might even be thought to be suffering from rabies. I have noticed that crows, which inflict severe loss upon toads at the spawning season, tear off the skin and turn it inside out so that they are not exposed to its effects while eating the flesh; nor do they eventually swallow the skin, but leave it to rot upon the ground. The stories and legends of toads spitting venom are fabrications.

The limbs are relatively short, especially the hinder, and the powers of jumping are correspondingly feeble. The hind toes are provided with partial webs only, in accordance with the terrestrial habit; the creatures only resort to the water at the breeding season and swim but little. They hide themselves in holes, and among stones, under tree trunks and in other cool places during the day and come out chiefly in the evening in search of food. They are most useful in gardens in keeping down slugs, snails, flies, beetles, wood-lice, caterpillars and other insect larvæ and they also devour earthworms. The prey is seized by the tongue in the same manner as with the frog. The jaws of a toad differ from those of a frog in being quite destitute of teeth, and the tip of the tongue is scarcely or not at all bilobed. The internal anatomy of the two is in general very similar, one of the most striking differences being in the position of the anterior abdominal vein. This vessel in toads lies dorsal to the muscles forming the ventral wall of the abdomen, whereas in frogs it runs in the substance of the muscles themselves. The toad, like the frog, periodically sloughs the outer layer of the skin; the process takes place about five times each season; the cast-off layer is swallowed.

The breeding season of the toad is rather later in the year than that of the frog. The spawn, instead of being deposited in masses, is in ropes, three or four feet in length, each rope consists of a double row of eggs arranged alternately on either side. The mucous envelopes of the eggs are rather more transparent than in frogs' spawn. The hind legs of the female are employed in thrusting the spawn-rope backward away from the body as the eggs are being laid. The general course of development does not differ in any important particular from that already described: the young tadpoles are, however, rather smaller and of a more intense black colour.

Toads are said to attain a very great age. I am not aware of any exact trustworthy records of the length of life, but have no hesitation in asserting that the reports of toads found entombed in solid chalk and other rocks have originated in imperfect observation of the facts. The absence of ribs makes it possible for a large toad to squeeze itself through a very narrow crevice. At the approach of winter the animals retreat into deep holes or other sheltered spots, and become torpid through the cold weather.

(3) The Natterjack Toad, *Bufo calamita,* is of rarer

occurrence than the common toad but has been found in many parts of England and also in Scotland. It frequents sandy heaths, especially such as are swampy in places. I have found it among the heather on the commons in south-west Surrey. It is more brightly coloured than the common toad and may be at once distinguished by the possession of a yellow line down the middle of the back. It also has several blotches of pinkish-red colour upon the back and sides, especially in the anterior regions and numerous small pale yellow specks. The eyes are very prominent and the skin closely covered with warts. The male possesses a large median vocal sac beneath the floor of the mouth. On opening the skin the surface of the sac is seen to be deeply pigmented. There are two slit-like apertures in the floor of the mouth by which air can be forced into the sac. The voice of this toad, when reinforced by this resonator, is extraordinarily powerful. A number in chorus can be heard for about half a mile across open country. If one toad begins to croak the cry is quickly taken up by all its neighbours. I have also noticed that the "laugh" of the great green woodpecker will often start them croaking. The webbing of the toes is even less pronounced than in the common toad. It is, for a toad, extraordinarily active, running with considerable speed, but seldom if ever jumping. It resorts to the water only in the breeding season, which occurs in May and June. The spawn is in ropes, resembling that of the common toad.

While the animals are in the water during the breeding season the lungs are maintained in a state of remarkable distension and greatly increase the apparent size of the toads. It is probable that at this period the lungs subserve a hydrostatic rather than a respiratory function, this last being now adequately performed by the wet skin and the mucous membrane of the mouth, perhaps assisted by the vocal sac. After leaving the water the toads appear to shrink in bulk owing to the return of the lungs to their normal size. If however a specimen is thrown back into the water, after spending a few days on land, it will at once inflate its lungs and "swell visibly."

Urodela (Tailed Amphibians).

The three species of British Newts or Efts closely resemble one another, so that their general characters and habits may be described collectively. They differ from

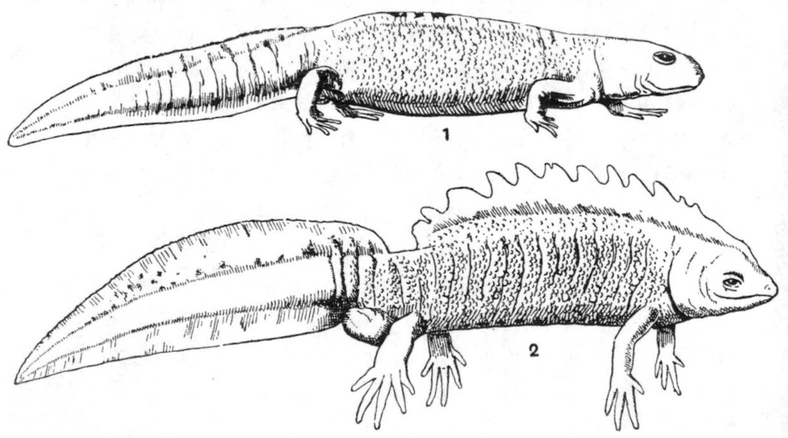

Fig. 50. *Molge cristata*, the Warty Eft. From Gadow.
1. Female. 2. Male at the breeding season with the frills well developed.

the frogs and toads in possessing a slender body and long tail with which they swim rapidly, and which retains the lateral line sense organs throughout life. Their legs are short, front and hind about equal in length, and of minor importance in aquatic locomotion. In general appearance they resemble lizards, which belong to an entirely different class, the Reptiles, and are sometimes mistaken for them. They may however be at once distinguished by the absence of claws from the toes—no British, and hardly any other Amphibians possess any nails or claws. Moreover the skin of a newt is soft and more or less moist, while that of lizards is hard and dry, and the tail of a newt is flattened from side to side so as to be narrowly elliptical in section, but a lizard's tail is circular in such view. Also the tongue of a newt is white and fleshy, that of a lizard black, thin and forked. The habits of the two are entirely different, for though newts often wander far from water they avoid intense heat and are sluggish in their movements, whereas lizards love to bask in the sun, and are exceedingly swift and agile.

All three species of newt are carnivorous, the food consisting of worms, small molluscs, insects, and occasionally small fish, tadpoles and even small individuals of their own species. Respiration is carried on in the same manner as in the frog. The breeding season is in the early summer. There is no union between the two sexes, but patches of spermatozoa are deposited by the males and then seized by the females and conveyed to the cloaca. The pale yellow eggs are laid singly on the edge of the leaf of some water weed, *Polygonum persicaria* usually, which is folded together by the hind feet of the female and retained thus by the sticky covering of the egg itself. Mr Douglas English tells me that he has observed a smooth newt swimming about with her egg "palmed" in the hind foot until she found a suitable leaf, which in this case proved to be of Starwort (*Callitriche aquatica*). From the eggs emerge tadpoles not very different from those of the frog: in the course of development however the external gills become far more complicated. The anterior limbs are developed earlier than the posterior. The metamorphosis takes place in late autumn. At the close of the breeding season most newts leave the water and pass the rest of the summer on land. During the daytime they remain hidden in holes and crevices but come out in the evening in search of food. During their life in the water they frequently come ashore at dusk and return to the water again before sunrise. The young newts on reaching the adult form leave the water and do not return to it, at any rate for any lengthened stay, until they have reached sexual maturity in their third or fourth year. As in the tailless amphibia, the skin is cast periodically, either in several pieces or whole, and is often immediately devoured. The males, at the breeding season, possess a well-marked dorsal crest continuing the line of the dorsal tail-fin towards the head.

Teeth are present on the bones of the upper and lower jaws and also upon the vomers and palatine bones in the roof of the mouth: they develop very early, preceding in time of appearance the bones upon which they are eventually carried. The radius and ulna in the front

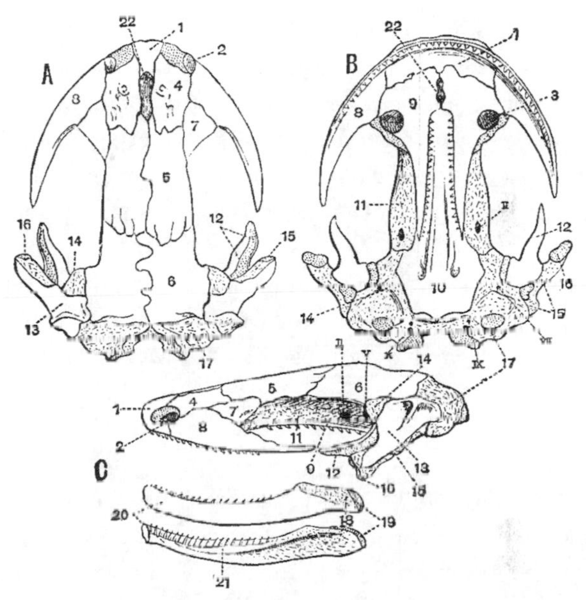

Fig. 51. A dorsal, B ventral, and C lateral views of the skull of a Newt, *Molge cristata* × 2½. After Parker.

The cartilage is dotted, the cartilage bones are marked with dots and dashes, the membrane bones are left white.

1. Premaxilla. 2. Anterior nares. 3. Posterior nares. 4. Nasal. 5. Frontal. 6. Parietal. 7. Prefrontal. 8. Maxilla. 9. Fused vomer and palatine. 10. Parasphenoid. 11. Orbito-sphenoid. 12. Pterygoid. 13. Squamosal. 14. Pro-otic region of fused exoccipital and pro-otic. 15. Quadrate. 16. Quadrate cartilage. 17. Exoccipital region of fused exoccipital and pro-otic. 18. Articular. 19. Articular cartilage. 20. Dentary. 21. Splenial. 22. Middle narial passage, a cleft in the cartilage of the snout filled with connective tissue. II. V. VII. IX. X. Foramina for the exit of cranial nerves.

limbs, and the tibia and fibula in the hind remain respectively distinct from one another, and in the skull the frontal bones are not united with the parietals. In these particulars the skeleton is of more primitive type than that of the tailless genera. The vertebræ have transverse outgrowths from the centrum in addition to the transverse processes springing from the neural arch: the neural spines are well developed: the centra concave posteriorly. Short ribs, not embracing the body, are present and articulate both with the neural and central transverse processes.

Newts possess a remarkable power, not exhibited by frogs and toads, of repairing injury: entire limbs are grown again from the stumps left after accidental or other amputation.

(1) The Crested Newt, or Great Warty Newt (*Molge cristata, Triton cristatus*). Length about 6 inches; colour black, or olive-brown with black spots dorsally, sides speckled with white; ventral surface orange-yellow with black spots and marblings. Surface of skin rough, warty. Toes yellow with black rings. The dorsal crest of the male is serrated, his head marbled black and white, and tail marked with a silvery-grey lateral band in the breeding season. In the skeleton there is no ligamentous or bony arch covering the frontal and squamosal bones. Common in all parts of the country.

(2) The Common Newt, or Smooth Newt (*Molge vulgaris, Lissotriton punctatus*). Length 3 or 4 inches; colour olive-brown with darker blotches, head with five longitudinal dark stripes; ventral surface yellowish with dark brown or black mottlings. The dorsal crest of the male is festooned and the hind toes lobed at the breeding season. The lower edge of the tail is reddish with a blue-black margin and interrupted by black bars in the male,

in the female it is uniformly orange. The eggs are said to be laid in rows occasionally. The frontal bone is connected with the squamosal by a ligamentous arch. Abundant in all clear ponds and ditches.

(3) The Palmate Newt or Webbed Newt (*Molge palmata, Lissotriton palmipes*). Length 3 inches or less; colour brown or olive with small dark spots dorsally; ventral surface with orange-yellow median stripes with a few black dots. Dorsal crest of male not serrated or festooned: in colour, as also the hind feet, black. Ventral edge of tail orange in female, dark grey in male. Hind feet of male webbed during breeding season. The tail of the male terminates in a slender filament, devoid of fin-like expansions, which is used in a prehensile manner, being twisted round stems of aquatic plants. This is the rarest of our newts, but has been reported from many different parts of England and Scotland. Apparently it chiefly frequents moorland and heathy swamps or ditches. I find it in S.-W. Surrey in the same localities as the Natterjack.

CHAPTER IX.

SOME COMMON INTERNAL PARASITES OF DOMESTIC ANIMALS[1].

In the previous chapters frequent reference has been made to the parasites usually associated with the several animals that have been dealt with. We will now discuss a few of those which are of general interest, inasmuch as they affect domestic animals and are therefore of economic importance.

Protozoa (Unicellular Animals).

A parasite of microscopic size that at times causes serious disease and loss among rabbits is the Protozoon *Coccidium oviforme*. As the specific name implies the organisms are of an oval shape: their length is 40—49μ, their breadth 22—28μ. They occur in enormous numbers in the liver and bile-passages of infected animals. The

[1] For detailed treatment of this subject *vide Traité de Zool., Medic. et Agric.*, 2nd ed. Paris; *United States' Dep. Agric. Bureau of Anim. Indust.*, Bulls. IX. XII.; *Zoologie Agricole*, Railliet; *Compendium der Helminthologie*, von Linstow, Hannover, 1878; *Zeit. f. d. ges. Naturw. Giebel*, LIV. 1881; Riehm., *Animal Parasites and Messmates*, Internat. Sci. Ser. XIX. 1876.

symptoms of their presence are general weakness, anæmia, diarrhœa, enlarged liver, and "pot-bellied" appearance. In two or three months the victims waste away and die. The post-mortem appearance of the liver exhibits a number of white specks and blotches varying from the size of a pin's head to that of a hazel-nut. These nodules contain masses of encysted *Coccidia*. The surrounding lobules of the liver are frequently more or less atrophied. A stage of the life history of the parasite is passed in damp earth, or water or other suitably moist medium to which access is gained *viâ* the droppings of its host. It is therefore important to isolate, or destroy, infected individuals, to burn their litter, and otherwise cleanse and disinfect their hutches, troughs, etc. Dry food, clean drinking-water and general cleanliness in the surroundings are the best precautions against the disease among domesticated rabbits. Guinea-pigs are known to suffer from the same complaint, which has also been recorded in man himself. In this case it is probably contracted by drinking dirty or unfiltered water or by eating raw green food, salads, etc., which have not been properly washed.

Trematoda.

The Trematodes are Flat-worms devoid of external cilia, being invested by hard cuticle. They adhere to their hosts by means of suckers. The mouth is anterior, and leads into a muscular pharynx, which by alternate contraction and expansion pumps the juices of the host into the alimentary canal of the worm. The alimentary canal itself forks into two main branches, right and left, and these

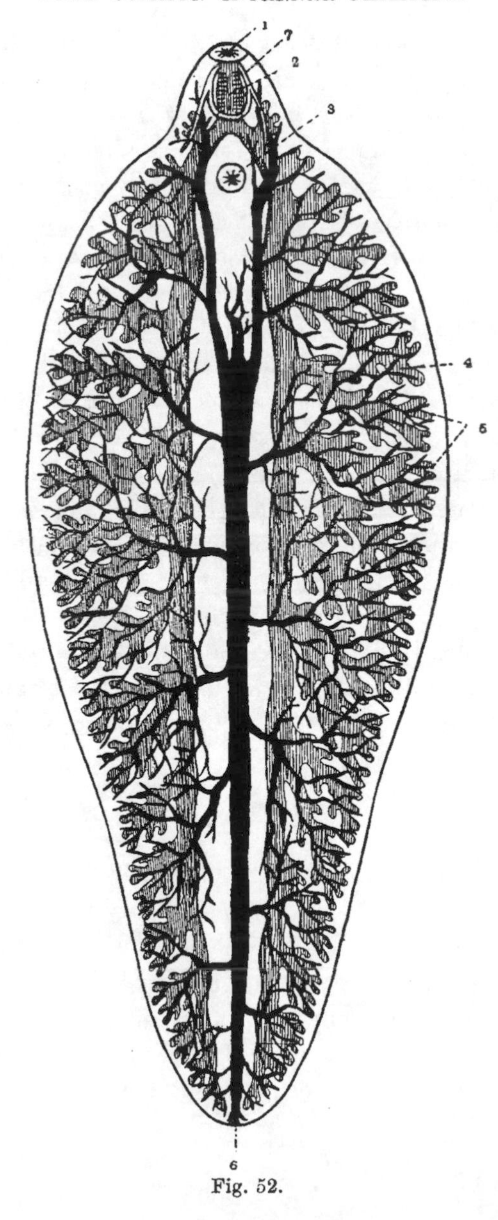

Fig. 52.

Fig. 52. Diagram of digestive and excretory system of *Distomum hepaticum* × about 8. From Leuckart.

1. Mouth. 2. Pharynx. 3. Reproductive pore. 4. Branch of alimentary canal. 5. Branches of excretory system. 6. External opening of excretory system. 7. Nerve-ring.

in turn may give off numerous ramifications extending through the body. There is no anus. The reproductive organs consist of a simple ovary, a pair of testes and numerous accessory glands: the entire genital system is, relative to the size of the animals, enormous, and a very large number of eggs is produced by each individual. With very few exceptions Trematodes are hermaphrodite, and some are undoubtedly capable of self-fertilisation.

The most notorious of all the Trematodes is the Liver-fluke, *Distomum hepaticum*[1], the cause of the disease known as "sheep-rot." The adult fluke is about an inch and a half long and is found in the bile-ducts and liver of the sheep and other domestic animals of similar habits. It has been known to occur in man. The eggs are deposited in the bile-ducts of the host and pass out with the fæces. They are protected by thick chitinous shells, and are about 0·13 mm. long and 0·08 mm. broad. Development of the embryo has already begun within the shell before the egg gets free from the body of the host, but it proceeds no further unless there be certain conditions, namely, moisture and a fairly high temperature, in the spot where the egg chances to fall. The optimum temperature has been found to be from 70°—75° F. Given the proper conditions there issues from the egg-shell, which opens by a circular lid at

[1] Thomas, *Q.J.M.S.* XXIII. 1883, and *Roy. Agric. Soc. Journ.* (2) 1881, 1882, 1883.

one end, a small ciliated larva about 0·15 mm. long and 0·04 mm. broad. The larva is provided with a double eye-spot, and at its anterior end has a central papilla which can be elongated to form a finely pointed awl. By means of its cilia the creature swims in the water and searches for a small water-snail, *Limnæa truncatula*. In our own country a marked preference is shown for this particular species of gastropod, though other species may be also attacked. Elsewhere however the embryo has accommodated itself to faunal differences. For example, Lutz[1] states that in the Hawaiian Islands, where there have been severe outbreaks of "rot," the intermediate host is *Limnæa cahuensis*. If the larva fails to find a snail within about eight hours it dies: on the other hand, if successful in the quest it proceeds at once to bore a passage into the soft tissues of the mollusc. This it achieves by elongating its awl and spinning round about its long axis by means of its cilia: it becomes for the nonce an animated drill. Mere penetration into the snail is not necessarily crowned with future success. In the tougher, more muscular portions of the body the larvæ soon perish. But in the pulmonary chamber and other highly vascular situations the larva, having no further need of organs of active locomotion, throws off its ciliated covering and becomes a mere sac. At the same time the eye-spots, being no longer of service, separate and eventually disappear. In this stage the organism is known as the *sporocyst*. Under favourable conditions it grows rapidly and in two or three weeks attains a length of about 0·6 mm.

[1] *Centrabl. f. Bakter. u. Parasitenk.* XI. 1892.

It is clear that the chances of any one larva meeting with a *L. truncatula* and even then of boring a way into a suitable tissue are very few. The infant mortality among flukes must be enormous. In the life-history of those few which are favoured by fortune there now occurs a series of events which to a large extent compensate for previous losses, and increase the prospects for the future, where, as will be seen, the odds are all against the fluke. Occasionally a sporocyst divides into two sporocysts. In any case each sporocyst produces within it, from what may be regarded as larval eggs, a number of secondary larvæ termed *rediæ*. A redia exhibits a decided step forward towards the adult condition; it possesses a mouth and pharynx, and a short, but unbranched, digestive sac. The rediæ are produced in succession and may be seen in all stages of development within the cavity of the sporocyst. When about 0·25 mm. long they become active and break out from the sporocyst into the tissue of the snail. By means of muscular contractions, aided by two blunt "feet" near the hinder end, they force their way into, generally, the liver, eating the tissues as they proceed and, if numerous, destroying the snail. Lutz mentions that in the Hawaiian epidemic there was a great mortality among *L. cahuensis* and that as many as two hundred rediæ were found in a single individual. The rediæ may attain a length of over 1 mm. A further multiplication of individuals derived from one successful larva now ensues. Each redia produces, after the same fashion as the sporocyst before it, a number of daughter rediæ which escape from their parent by a special aperture placed laterally just behind a circular thickening of muscle known as the collar. If the temperature conditions be favourable several generations of rediæ are produced. Eventually, in the autumn, all the rediæ give birth to other forms known as the *cercariæ*. Of these each redia produces from fifteen to twenty. A cercaria is about 0·28 mm. long, and is rather like a tadpole in outline. The anterior part is flat and heart-shaped, and possesses two suckers, a mouth, pharynx and forked digestive tract; the posterior portion, which is about double the length of the anterior, consists of a very muscular and contractile tail. In this stage the parasite makes its way out of the body of the snail and swims in the water or wriggles along among the leaves of plants. Soon it comes to rest upon a blade of grass or other similar object, shakes off its tail, and secretes over itself a white calcareous case. In this condition the cercaria can remain, encysted upon the herbage, for some time. It develops no further except it pass into the stomach of a sheep or some other herbivore. Here the wall of the cyst is dissolved by the gastric juice, and the liberated cercaria makes its way up the bile-duct into the liver where it rapidly grows and becomes sexually mature in about six weeks. Within the liver they can certainly live for three years and probably longer.

It is to be noted that *L. truncatula*, though an aquatic snail, frequently leaves the water and is capable of withstanding considerable drought. Hence the infection may linger in a pasture for a surprisingly long time. The fact that sheep suffer from "rot" far more than cows is due to the encystment of the cercaria occurring upon the lower leaves of grass and other plants. The cysts are thus more frequently swallowed by the close-cropping sheep than by cows, which wrench off merely the upper portions of the leaves at each bite. Watercress from infected districts, especially if not thoroughly washed in salt water, is a possible source of infection occurring to man.

Symptoms. The symptoms and course of the disease in sheep are as follows. The infected animal at first appears to fatten. This appearance is not caused by any real deposit of adipose tissue, but is due merely to serous infiltration into the subcutaneous connective tissue. The appetite then fails, great thirst is manifested and rumination becomes irregular. At the same time the skin and mucous membranes of the mouth and nostrils become of a whitish-yellow tint. Gradually the animal gets weaker, and in from two to six months the disease proves fatal if the parasites are present in any quantity. In severe cases one sheep may have as many as a thousand flukes in its liver.

Precautions. The precautions to be observed are (1) to avoid pasturing the flocks upon wet land or such as is liable to be inundated by floods. In such situations *L. truncatula* is likely to thrive in abundance. (2) To apply top dressings of lime and of salt to the pasture: by this means both the snails and any cercariæ that may be present are destroyed. Salt, even in comparatively small quantities, is peculiarly fatal to the fluke in all stages of its existence. (3) To supply dry food and blocks of salt for the sheep to lick. (4) To destroy the droppings of infected sheep, for it is in these that the eggs of the parasite are present; and to slaughter badly infected individuals.

It will not be out of place to draw attention to the losses suffered in past years by the ravages of the Liver-fluke, and to the value of exact zoological knowledge in enabling us to combat the disease. In 1830 sheep to the value of four millions sterling were lost by "rot" in England alone. In 1879–80 about double that amount perished from the same cause. In Ireland during 1862 sixty per cent. of the flocks were fatally affected. Of recent years, since Thomas's researches have made known the life-history of this pest, outbreaks have not been so serious.

Of other Trematodes we will mention but three, which are parasite in birds. *Monostoma mutabile* is found in the suborbital sinuses, nasal cavities, trachea, air-sacs, and intestines of aquatic birds and occasionally of others. Its redia stage is almost certainly passed in some small snail.

M. flavum occurs in the trachea and œsophagus of duck and other lamellirostrate birds. The intermediate host is the snail *Planorbis corneus.*

Notocotyle verrucosum becomes sexually mature in the intestine of fowls, ducks, geese and other birds. Its intermediate host is not known with certainty, but is probably a snail of the genus *Planorbis* or *Limnæa.*

Cestodes (Tapeworms).

These flat-worms though outwardly exhibiting an apparent segmentation are usually, for reasons which are beyond the scope of the present work, regarded as unsegmented worms. The "head" is armed with suckers, or hooks, or both appliances, for secure attachment. There is no alimentary canal, but the digested food of the host is absorbed direct through the general surface. Neither are there any special organs of locomotion. The generative organs are of enormous relative proportions, and are present, both male and female, in each joint (proglottis). The sexes, however, are not ripe simultaneously but each proglottis functions first as male and subsequently as female. The prodigious fecundity of these parasites is correlated with the enormous mortality of the embryos. They usually make no effort to find either an intermediate or final host. Success is a mere matter of luck. They are of economic interest inasmuch as not only do they occur as parasites in man, but also, either as bladder-worms, or when mature, in many of our domesticated animals, causing disease of greater or less severity. The species that most commonly come under observation are those that infest the alimentary canal of the dog, and make their presence known by the escape *per anum* of white flattened, oblong, "joints" (proglottides), possessed of feeble powers of locomotion. Other symptoms exhibited by the dog are a harsh, staring coat, foul breath and occasional scraping of the rump along the ground. No less than six species of the genus *Tænia* are known to occur in the dog, in addition to some three or four species of other genera. The most usual is known as *Dipylidium caninum.* The eggs of the worm are set free by the decay of the wall of the detached ripe proglottis and are devoured by certain

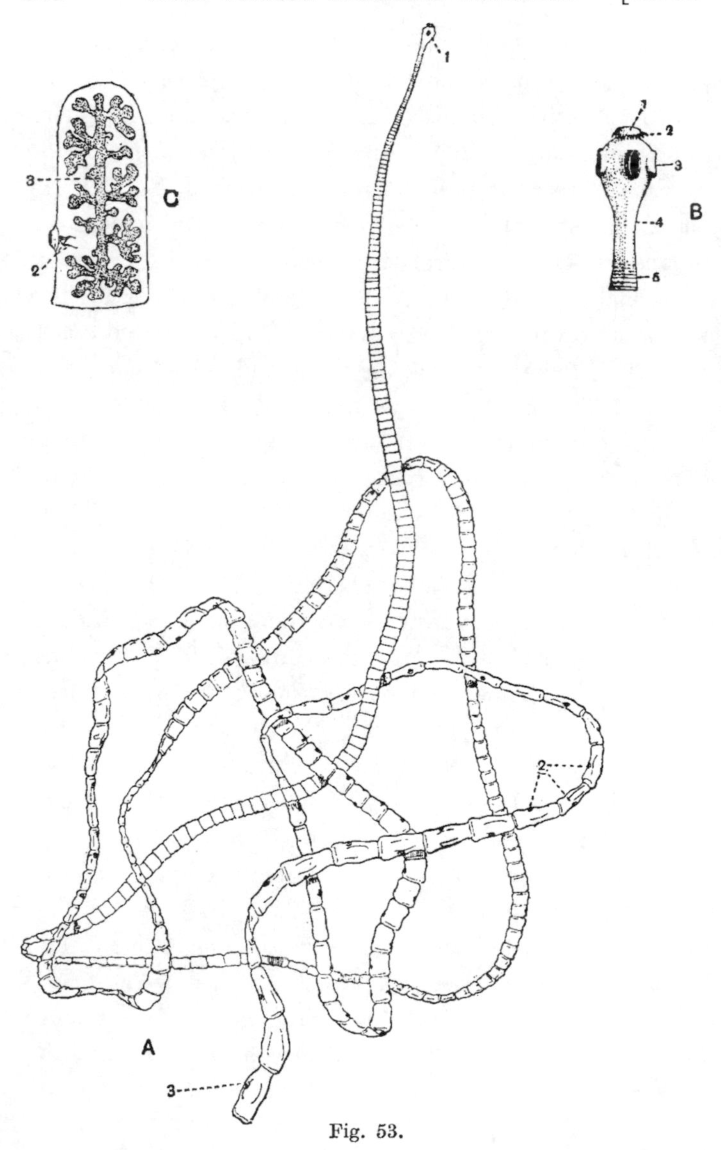

Fig. 53.

Fig. 53. *Tænia solium.* Slightly magnified.

A. Entire worm showing head and proglottides. 1. Sucker on head. 2. Genital pores. 3. Ripe proglottis.

B. Head. 1. Rostellum. 2. Hooks. 3. Suckers. 4. Neck. 5. Commencement of strobilization.

C. Ripe proglottis broken off from worm. 2. Remains of vas deferens and oviduct. 3. Branched uterus crowded with eggs.

external parasites of the dog, such as the flea (*Pulex serraticeps*) and louse (*Trichodectes latus*), and give rise to the bladder-worm in these intermediate hosts. Beyond this stage it does not progress except it be devoured by dog, fox, wolf, etc.; in which event the bladder disappears and the "head" of the worm attaching itself to the lining of the intestine of the new host buds off a long series of "joints," each of which in due course after sexual maturity becomes crowded with fertile eggs. The success of this parasite then depends upon the dog nibbling and biting at the fleas or lice which are irritating him, especially when these happen to be in the region of the anus, and the hair of the dog matted with fæces.

In dissecting a rabbit one may often meet with small bladder-worms attached singly or in clusters to the mesentery; this form is known as *Cysticercus pisiformis*, and is, as its name indicates, about the size of a pea: it gives rise in the dog to the tape-worm *Tænia serrata*. The cystic stage of *Tænia cœnurus*, another species found in the dog, is *Cœnurus cerebralis*, which occurs in the brain and spinal cord of sheep and horses and causes the complaint know as "gid" or "staggers." This bladder-worm may be as large as a walnut and contains numerous "heads" each capable of producing a complete tapeworm.

In some species the bladder-worm may surpass even these dimensions and become as large as a swan's egg or even as a man's head (*Echinococcus polymorphus*).

It is most remarkable how these parasites have succeeded in adapting themselves to the conditions necessary to their perpetuation. Indeed a special providence seems to have cared for them. It will doubtless have been noted that in the above-mentioned instances the intermediate host is one that in the ordinary course of wild nature is likely to fall a victim to the carnivorous animal which is the final host of the tapeworm. One of the most striking examples of the singular adaptation is found in the life-history of the tapeworm *Bothriocephalus cordatus*. The bladder-worm stage is here passed in fish; Icelanders are in the habit of giving their dogs fish to eat and in the dogs the worm reaches maturity. It is probable that the natural host of this worm is some aquatic carnivore, such as the seal.

It may not be out of place, seeing how frequently domestic dogs are affected with these unpleasant parasites, to emphasise the importance of never allowing a dog to *eat* mice or rats, or the entrails of rabbits, sheep, and other animals in which the intermediate stages of the worm are passed. It is only by eating food of this character that it is possible for a dog to "catch" tapeworm. It is also important to keep the animals well groomed, free from external parasites and, in the case of long-haired dogs especially, to cut away hair that may have got soiled *circum anum*.

Upwards of two dozen different species of tapeworms

belonging to the genera *Tænia*, *Drepanidotænia*, *Dicranotænia*, *Bothriotænia*, *Echinocotyle* and *Davainea* are known to infect poultry, including geese, pigeons and pheasants as well as ducks and fowls in this term. The intermediate hosts are not known in every instance, but where they have been determined they are always the natural food of the final host. Thus the tapeworms of the duck and goose pass the bladder-worm stage in small fresh-water crustaceans such as *Gammarus*, *Cyclops*, and *Cypris*. Those of the fowl, pheasant and pigeon in earthworms (*Allolobophora fœtida*), slugs (*Limax*), house-flies and other insects—the pupæ of ants being suspected of harbouring one species (*Davainea friedbergeri*) that becomes mature in the pheasant.

Mice and rats are liable to be infected by a small tapeworm (*Hymenolepis diminuta*), whose bladder-worm stage is found in various insects, but especially in the meal moth, *Asopia* (*Pyralis*) *farinalis*. The habits of this moth render it peculiarly likely to come within reach of the above rodents.

It is somewhat surprising to find that purely herbivorous animals are not exempt from this class of entozoon. About a dozen different species of the genus *Moniézia* have been described from the intestines of sheep, goats, oxen and other ruminants. Others of like habits belong to the genera *Thysanosoma* and *Stilesia*. Similarly from the intestines of rabbits, hares, horses and asses numerous tapeworms, some small, others attaining a length of two or three feet have been taken. Those of the horse are of the genus *Anoplocephala*, of the rabbit

Ctenotænia and *Andrya*. In none of these instances is the life-history known and much remains to be done in elucidating the mode of infection.

A tapeworm can only be dislodged by resorting to purgatives. The animal should be denied all food for a day, and then areca nut or male fern administered. Thymol is also a valuable remedy, but care should be taken to avoid administering it in company with any solvent such as oils (or, in the case of man, alcohol), lest it should be absorbed into the system.

Nematoda (Thread- or Round-Worms).

The majority of these remarkably hard, unsegmented worms are of minute, and often microscopic, size, but some species attain a length of several feet. Their smooth surface and finely tapering body enable them to insinuate themselves into the tissues of animals and plants, and account for their extraordinary prevalence. They are more generally known by the diseases they produce in animals and in plants than by their actual appearance. In this country, of those parasitic in man, *Ascaris lumbricoides*, which occurs in the intestine more commonly in children than in adults, is the best known. The eggs develop in water or damp earth and liberate their embryos when re-introduced into the human alimentary canal. Another species of this genus, *A. megalocephala*, occurs in horses and asses, and may sometimes be seen, being a large worm about 10 inches in length, in the droppings of these animals. The complete life-history of this species is not known with certainty.

A species more important, from the effects produced by it in man, is the microscopic *Trichina spiralis*, the cause of the disease trichinosis. This form is parasitic throughout

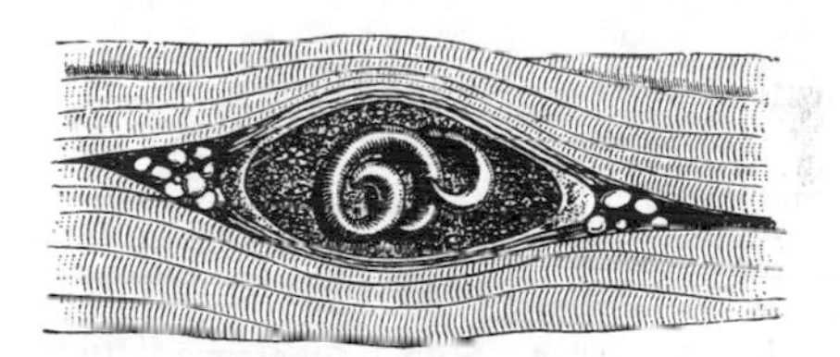

Fig. 54. *Trichina spiralis*, encysted amongst muscular fibres. Highly magnified. After Leuckart.

its entire life-history and has no free stage of existence, such as is not infrequent among other Nematodes. The disease is contracted by eating "underdone" pork infested by the encysted miniature *Trichinæ* which have lodged themselves in the muscle fibres of the pig, where they are capable of remaining alive for many years. The larvæ are liberated from their cysts by the digestive fluids of the unfortunate human being and in about three days become sexually mature. They then resort to the tubular glands of the small intestine[1]. Here a brood of larvæ, amounting to as many as 1,000 individuals from each female, is, in about another four days, viviparously produced. These bore their way through the lining of the intestine into the blood vessels of their host, occasioning severe symptoms, and at times even death. If death is not caused the larvæ are swept along by the blood-stream, and eventually, aided by their own wriggling movements, pierce their way

[1] Geisse, *Münchener med. Wochensch.* XLII. 1895; and *Zool. Centralbl.* IV. 1897.

into the intra-muscular connective tissue[1], and develop around themselves a lemon-shaped cyst which in course of time becomes calcareous, while the affected fibres undergo degeneration. The symptoms at this stage are in some respects similar to those of rheumatism. The natural host of *Trichina* is the common rat, which is a dirty and omnivorous feeder, and does not scruple to eat the dead bodies of its brethren, and thus the infection is passed on from generation to generation. Occasionally infected rats are devoured by pigs, and thus man is exposed to the risk of trichinosis.

In tropical countries there are numerous other diseases for which species of Nematodes are responsible. Intermediate hosts have been discovered in some cases, notably the mosquito as the introducer of *Filaria sanguis hominis*. In others there is no intermediary, but infection is direct. To one of these, *Uncinaria* (*Anchylostoma*) *duodenalis*, special interest is attached from the fact that it was introduced some seven or eight years ago into Dolcoath Mine, Cornwall. This misfortune was doubtless brought about by some miner returning infected from a tropical country. Dolcoath Mine is some 3,000 feet in depth and the temperature in it ranges from 75° F. to 90° F. The conditions are therefore well suited to the parasite. The male is from 8 to 11 mm., the female from 10 to 18 mm. long. They inhabit and become sexually mature in the first portions of the small intestine. The eggs develop to a certain stage within the parent and are evacuated with the fæces of the host. Incubation only takes place in a

[1] Hertwig and Graham, *Münchener med. Wochensch.* XLII. 1895.

semi-solid medium such as fæces, moist earth, etc.; the larvæ hatch in from one to four days according to the temperature. They are about 0·21 mm. long and 0·04 mm. thick, and are provided with three teeth. Growth is rapid and on the third day after hatching the skin is moulted. In about a week they attain a length of 0·56 mm. They then moult again and throw off their teeth, but may remain within the cast-off skin, which becomes calcareous. In this condition they can withstand prolonged desiccation, and get blown about by the wind. They are now capable of living free in water, especially if this be dirty. They probably gain access to man in drinking-water, or food contaminated by soiled hands. The adult stage is reached in some weeks after further moults. The worms inflict wounds upon the lining of the intestine, and may cause death. Their presence is indicated by anæmia of the patient and by the presence of the eggs in the fæces. At the present time the condition of the Dolcoath Mine has been greatly improved by forced ventilation, lowering the temperature, by free application of disinfectants to spots likely to harbour infection, and by the enforcement of sanitary rules.

A closely allied form, *Uncinaria trigonocephala*, occurs in the intestine of dogs and cats. The male is from 9 to 12 mm., the female from 9 to 21 mm. in length. Its life-history resembles that of the last species. After its second moult it can remain alive, though motionless, for many months in water. Domestic animals probably swallow it in this stage with their drinking-water. Another larger form often found in the same hosts is *Ascaris nuptax*.

probably acquired in the same way. Horses suffer from the attacks of a worm of similar habits, *Sclerostomum* (*Strongylus*) *equinum*. The males of this species vary in length from 18 to 35 mm., the females from 20 to 55 mm. When immature they inhabit the blood vessels of their host and may cause serious injuries. The disease produced is known as "colic." The adults make their way to the mucous membrane of the cæcum and when mature pass into the large intestine. The eggs are expelled with the fæces of the horse and develop in water, giving rise to minute larvæ which enter the horse with his drinking-water, and at once penetrate into the blood vessels of the small intestine.

A number of Nematodes inhabit the air passages of higher animals. Two species, *Strongylus filaria* and *St. rufescens*, occur in the trachea, bronchial tubes and air cells of sheep and goats. The former is from 3 to 10 cm., the latter from 1·8 to 3·5 cm. long. In neither case is the intermediate host known with certainty, though it has been determined that such is necessary.

The disease known as "gapes" in poultry and other birds is due to a small member of this order, *Syngamus trachealis*. The length of the male is from 2 to 6 mm., that of the female from 5 to 20 mm. They occur in the windpipe of the birds. Sexual union results in permanent adhesion, hence the adults appear to be double-headed. The eggs escape only by rupture of the body of the female, inside or outside of the body of the host. They develop in water and may even hatch. The free embryo is 0·28 mm. long and 0·01 mm. thick. It has however been shown by Ehlers that if eggs containing embryos are swallowed by birds, mature individuals are found in the trachea within a fortnight. Walker has further proved that the embryos remain alive within the alimentary canal of earthworms, and that birds which eat such worms in a short time suffer from "gapes." Pigeons and poultry also suffer from the attacks of *Heterakis maculosa* (*Ascaris columbæ*) and *H. papillosa.* These worms swarm in the intestines of the birds and may cause death. Both are of small size; the former from 16 to 34 mm., the latter from 7 to 15 mm. in length. The eggs develop in moisture, and the larvæ reach maturity in about three weeks. No intermediate host is required, the parasites entering the birds with the food.

A large number of plants are injuriously affected by these worms. "Clover-sickness," "beet-sickness," "ear-cockles" in wheat, and frequent diseases of the roots of tomatoes, cucumbers, cabbages and other plants are all directly due to these "eel-worms." Indeed there is hardly any plant that may be regarded as entirely immune from their ravages. The small species often found in stale paste and in vinegar, and known as "paste-worms" and "vinegar-eels," are members of this successful order of parasites.

As a group they are most difficult to combat, since most can withstand prolonged drying, even for years, with unimpaired vitality.

INDEX

Abraxas grossulariata 87
Acanthinula 212
Acicula lineata 237
Acroloxus lacustris 237
Adrenal bodies 277
Æolosoma gen. et 6 sp. 33
Æolosomatidæ family 33
Æschna et 6 sp. 117, 118
 cyanea, figure of metamorphosis 108, 109
Æschnidæ 112, 116–118
Agriolimax gen. et 2 sp. 245
Agrion et 3 sp. 121, 122
Agrionidæ 119–122
Alar cartilages 265, 266
Alary muscles 86
Allolobophora subgen. et 4 sp. 42, 43
 fœtida 315
Allurus syn. 41
Ammophila hirsuta 134, 153
Amœboid corpuscles 257
Amphipeplea glutinosa 238
Amphistoma subclavatum 291
Anagrus incarnatus 103
Anax imperator 117
Anchylostoma 318
Ancylus fluviatilis 237
Andrya 316
Anisolabis annulipes 93
 maritima 93
Anisopterides 112, 113 (fig. of wings)
Anodonta, action of chief muscles 171
 circulation 178–181
 contents of alimentary canal 174
 crystalline style 176–178
Anodonta cygnea 162–204
 development 184–196
 endurance of frost 196
 excretion 181
 food and digestion 173–178
 genus, definition of 203
 "liver", action of 175
 locomotion 169–172
 muscular impressions 165 (fig. 29), 166
 nervous system 182–184
 parasites of 197
 post-parasitic young 195
 rate of growth 163
 rate of progression 169
 reproduction 184 sq.
 respiration 178–181
 structure of shell 164
 turgescence of foot 172
Anopheles maculipennis 133
Antenna-comb of wasp 77
Antennæ, functions of in cockroach 87
Anura 292
Aplecta hypnorum 238
Apterygida albipennis 93
 arachidis 93
Arion gen. et 6 sp. 245
 ater 212, 229, 230
Ascaris 22
 columbæ 321
 lumbricoides 316
 megalocephala 316
 nuptax 319
Asopia farinalis 315
Aspidogaster 198
Assemania grayana 236

Astacobdella 69
Astacus fluviatilis 48
  torrentium 48
Atax bonzi } 197, 198
  ,, *crassipes* }
Atlas vertebra 269
Aulostoma 28, 29
Auriculidæ 239
Azeca tridens 243

Bacteria 2, 12, 13, 58
Balea perversa 244
Bed-bug 80
Bee, pollen brushes 78
Beetles, enemies of molluscs 231
Beet-sickness 321
Bimastus subgen. et 2 sp. 44
Bithynia gen. et 2 sp. 235
Black-beetle 71
Bladder-worms 311
Blatta 71 et seq.
  americana 92
  germanica 91
Blattodea 94
Blood, cockroach 85
  crayfish 58
  earthworm 9
  frog 257
  leech 29
  mussel 179
  snail 222
Bothriocephalus cordatus 314
Bothriotænia 315
Botryoidal tissue of leech 30
Bohemilla gen. et 1 sp. 35
Bojanus, L. H. 181 footnote
  organs of 181
Brachytron pratense 117
Brain, frog 270–272, fig. 46
Branchiobdella 69
Branchiostegite 56, 57
Branchiura gen. et 2 sp. 36
Bristles on legs of insects 79
British Amphibia 292
  Dragonflies 112–122
  Oligochætes, list of 33–47
  Orthoptera 92–98
  Newts 296–301
  Wasps 159–161
Bufo calamita 294–296
  vulgaris 292

Buzzing of insects 131–134

Calcareous sacs, frog 271, 274
Calciferous glands, earthworm 6
Calopteryx et 2 sp. 119
Carychium minimum 239
Cat, round worms of 319
Cattle, tapeworms of 315
Cercaria 308
Cestodes 310–316
Ceylon pearl fishery 200
Chætogaster gen. et 2 sp. 34
Chironomus, hæmoglobin in blood
  of 10
Chloragogen cells of earthworm
  11, 30
Circulation, cockroach 85, 86
  crayfish 58
  earthworm 9, 10
  frog 257–261
  mussel 178, 179
  snail 220–223
Clausilia 212; gen. et. 4 sp. 244
Cleansing apparatus of insects 76,
  78
Clepsidrina 92
Clepsine 29, 32
Clitellio gen. et 1 sp. 37
Clitellum of earthworm 15
Clover-sickness 321
Coccidium oviforme 302, 303
Cochlearia 232
Cochlicella 242
Cochlicopa lubrica 243
Cockroach 71–92
  circulation 85, 86
  cleansing apparatus 76
  digestion 80–82
  eggs of 90
  excretion 82
  foes 92
  food of 80
  growth of 91
  introduction to England 71
  legs 75, 76, 79, 80
  moultings 91
  movements of legs 80
  nervous system 86–88
  odour of 72
  parasites 92
  reproduction 89–91

Cockroach, respiration 83, 84
  sense of smell 87
  wings 72, 74, 75
Cocoon of earthworm 17, 18
  leech 30, 31
Cæcilioides acicula 243
Cœnurus cerebralis 313
Cordulegaster annulatus 117
Cordulia ænea 116
  Corduliinæ 115, 116
"Crabs' eyes" 68
Cranial nerves 273
Crayfish 48 et seq.
  appendages, use of 51 53
  development 65, 66
  digestion 53–56
  excretion 58
  figure of, fig. 7, p. 50
  foes 69
  food of 49, 53
  gastric ossicles 53–55
  gastroliths 68
  geographical distribution 70
  heart 58
  "liver" 55, 56
  moultings 66–68
  nervous system 58 et seq.
  otocyst 60 et seq.
  parasites 69, 70
  regeneration 65, 68, 69
  reproduction 65, 66
  respiration 56–58
  sense of smell 65
  supply to market 48, 49
Crested Newt 300
Crickets 96
  field 97
  house 98
  mole 98
  noise of 89
  wood 97
Criodrilinæ subfamily 41
Ctenotænia 316
Cuckoo 112
Currant Moth 87
Cyclas, locomotion 2, 15
Cyclops 315
Cyclostoma vide sub *Pomatias*
Cypris 315
Cysticercus pisiformis 313

Dart-sac, snail 229
Davainea 315
Dendrobæna subgen. et 3 sp. 43, 44
Denudation and earthworms 25
Dero gen. et 6 sp. 35
Dextral shells 205
Dicranotænia 315
Digestion, cockroach 80–82
  crayfish 53–56
  earthworm 5–9
  frog 251–257
  leech 28
  mussel 173–178
  snail 216 220
Dipylidium caninum 311
Discelis filaria 22
Dispersal of leeches 32
Distomum cirrigerum 70
  hepaticum 301, fig. 52, 310
  isostomum 70
  somateriæ 199
Dog 311
  round-worms of 319
Dolichopus 79
Dragonflies 99–122
  British, table of 112–122
  development 103, 107
  eggs 102, 103
  enemies 111
  food 101, 103, 104
  legs of 101
  "mask" 104
  mating 102
  metamorphosis 107–110
  respiration 85, 105, 106
  wings 100, 101, 113 (figure)
Dreissensia, genus, definition of 204
  polymorpha 176, 201, 204
Drepanidotænia 315

Ear-cockles 321
Earthworm and bacteria 2, 12, 13
  burrow 2, 3
  calcareous glands 6
  castings 24, 25
  chloragogen cells 11
  circulation 9
  clitellum 15
  cocoon 16, 17, 18
  coelomic fluid, properties of 12
  cuticle 1

Earthworm *cont.*
denuding agency 25
digestion 5
economics 23, 24
enemies of 20
excretion 10
food of 5, 6
habits 1, 2, 3
intelligence 2
locomotion 4
monstrosities 20
nephridia 10
odour of 13
parasites of 20
rain after 3, 10
recuperative powers 19
reproduction 15–19
respiration 9
sensitive to light 13, 14
sensitive to vibration 13
setæ 4, 5
slime 2
special senses 13
typhlosole 7, 8
"yellow cells" 11
Earthworms, archæologists, importance to 25
Earwigs 93, 94
Echinococcus polymorphus 314
Echinocotyle 315
Echinorhynchus polymorphus 70
Ectobia lapponica 95
livida 95
panzeri 95
Edible Frog 292
Eel-worms 321
Efts 296–301
Eggs of cockroach 90, 91
crayfish 66
dragonfly 102, 103
frog 282–284
leech 30
newt 297, 298
slug (*Arion ater*) 230
snail 229, (*Helix aspersa*) 230
wasp 141
Eisenia gen. et 3 sp. 42
Eiseniella gen. et 2 sp. 41
Ena gen. et 2 sp. 243
Enallagma cyathigerum 122
Enchytræidæ family 39
Enchytræus gen. et 2 sp. 39, 40
Endamœba blattæ 92
Enterochlorophyll 219
Epiphragm 210–212
Erythromma naias 120
Escargotières 233
Euconulus fulvus 240
Eumenidæ 125
Evania 92
Excretion, cockroach 82
crayfish 58
dragonfly 110
earthworm 10–13
frog 275–277
leech 30
mussel 181
snail 223
wasp 144, 155

Filaria rhytipleuritis 92
Filaria sanguis hominis 318
Flea 313
Flèche tricuspide 176
Forficula auricularia 93
pubescens 93
Forficularia 93
Fossores 123
Fresh-water Lamellibranchs, list of British 202–204
Fridericia gen. et 2 sp. 40
Frog 246 et seq.
circulation 257–261
colour 246
croaking 282
development 284 et seq.
digestion 251 et seq.
eggs 282–284
enemies 289
excretory system 275–277
eyeball 248
used in swallowing 253
fat-bodies 256
food of 251
hibernation 281
limbs 249
liver of 254–256
nervous system 267–274
nostrils 265, 266
parasites 289
reproduction 277 sq.
respiration 261–267

Frog *cont.*
sexual differences 279–282, fig. 48
skeleton 267–270, fig. 45
skin 246, 247
tongue 251, 252

Gall-flies 123
Gammarus 315
"Gapes" 320
Gastric ossicles of crayfish 53–55
Gastroliths 68
Gastropods, table of British land and fresh-water 235 245
Geniohyoid muscle 265
Geomalacus maculosus 245
"Gid" 313
Gills of tadpole 286 288
Glochidium 188–192, figs. 34–36
Glomerulus 275
Glossoscolecidæ family 40
Glottis 262, 266
Glow-worm 231
Glycogen 255
Gomphocerus 95, 96
Gomphus vulgatissimus 112, 117
Gordius 92
Grasshoppers, British 95
chirrup 88
Great Warty Newt 300
Green glands 58
Green Locust 96
Gregarina blattarum 92
Gregarines 69
Gryllodea 96, 97
Gryllotalpa gryllotalpa 98
Gryllus campestris 97
domesticus 97
Guinea-pig 303

Hæmatin 55
Hæmocyanin 57, 222
Hæmoglobin 9, 222, 257
Haplotaxidæ family 40
Haplotaxis gen. et 1 sp. 40
Heart, cockroach 85, 86
crayfish 58
frog 258–260, 274
mussel 179
snail 220
worm 9

Helicella 241, 242
Helicidæ 240–242
Helicigona 241
Helicodonta 242
Helix, blood of 222
circulation 220–223
enemies and parasites 230
excretion 223
eyes 223
heart-beat, rate of 220
hibernation 210
length of life 209
"liver" 218, 219
locomotion 213–216
rate of growth of shell 213
rate of progression 208
reproduction 226–230
respiration 220–222
species of 240–242
Helix aspersa 205, 234
eggs 230
Helix pomatia, shell of 205–210
Helodrilus gen. et 3 subgen. 42–44
Heterakis maculosa 321
papillosa 321
Hibernation of frog 281
snail 210
wasp 139
Hirudinidæ 26
Homing of bumble-bees 153
of wasps 152–154
Hornet 159
Horse, round-worms of 316, 320
tapeworms of 315
Horseleech 28, 31
Hydrobia vide sub *Paludestrina*
Hydrocharis 198
Hygrocrocis intestinalis 92
Hygromia 241, 242
Hyoid 264

Ichneumon-flies 123
Ilium 269
Intestine of tadpole 286
Ischnura et 2 sp. 121

Jaminia gen. et 4 sp. 243

Keber's organs 181, 198
valve 172, 179
Kidney 275, 278, fig. 47

Kreidl, experiments on Crustaceans 60 et seq.

Labia minor 93
Labidura riparia 93
Lateral line 287
Leech, blood of 29
  digestive system 27, 28
  dispersal 32
  excretion 30
  eyes of 29, 30
  farms 26
  food 27, 31, 32
  locomotion 26
  nervous system 29, 30
  reproduction 30
  respiration 29
Leeches 26 et seq.
Length of life of molluscs 230
Lepidoptera, fore and hind wings of 127
Leptophyes punctatissima 96
Lestes et 2 sp. 119, 120
Leucorrhinia dubia 114
Libellula et 3 sp. 114
Libellulidæ 112–116
Limax gen. et 3 sp. 244, 245
  flavus 212
  food of 216
  host of tapeworm 315
  maximus 212
  respiratory pigments 223
  sense of smell 224
Limnæa gen. et 7 sp. 237, 238
  auricularia 229
  cahuensis 306, 307
  eyes 224
  haemoglobin in 222
  stagnalis, Semper's researches on growth of 209
  subaquatic locomotion 214, 215
  truncatula 306–309
Limnodrilus gen. et 2 sp. 37
Lissotriton palmipes 301
  punctatus 300
Lithobius 21
Liver, crayfish 55, 56
  frog 254–257
  mussel 177
  snail 219
Liver-fluke 304, fig. 52, 310

Locust, noise of 89
Locusta viridissima 97
Locustodea 96
Lophomonas blattarum 92
  striata 92
  sulcata 92
Louse 313
Lumbricidæ, family 41
  table of British 47
Lumbricillus gen. et 1 sp. 39
Lumbriculidæ family 38
Lumbriculus gen. et 1 sp. 38
Lumbricus gen. et 5 sp. 45
Lymphatic system, frog 260, 261

Malpighian capsule 275
  tubules 82
Margaritifera vulgaris 200
Marionina gen. et 2 sp. 39
Meal-moth 315
Meconema varium 96
Metamorphosis, dragonfly 107–110
  frog 288, 289
  newt 298
Metæcus paradoxus 157–159
Mice, tapeworms of 315
Micrococcus conchivorus 197
Milax gen. et 2 sp. 244
Molge cristata 296, fig. 50, 299, fig. 51, 300
  palmata 301
  vulgaris 300
Moniézia 315
Monocystis 23
Monostoma flavum 310
  mutabile 310
Monstrosities, earthworm 20
Mosquito 318
  stridulating apparatus 132, 133
Moulting, cockroach 91
  crayfish 66–68
Mucous threads of molluscs 214–216
Mussel 162–204, vide sub *Anodonta*
Mylohyoid muscle 265
Mytilus edulis, pearls in 198, 199

Naididæ family 34
Nais gen. et 2 sp. 34
Natterjack Toad 251, 294–296, 301

Nematodes 21, 22, 92, 291, 292, 316–321
Nemobius sylvestris 97
Nephelis 29, 31
Nephridia of earthworm 10
Nephrostomes, frog 276
Neritina fluviatilis 236
Nervous system, cockroach 86–88
crayfish 58–65
frog 267–274
mussel 182–184
snail 223
Newts 296 301, figs. 50, 51
Notochord 267
Notocotyle verrucosum 310
Nyctotherus 92, 290

Octolasium gen. et 1 sp. 44, 45
Ocypus 21
Odonata 99
Omohyoid muscle 264
Opalina 290
Orthetrum et 2 sp. 115
Orthoptera, British 92 et seq.
Otoconia, snail 225
Otocyst, *Anodonta* 182, 183
of crayfish 60 et seq.
snail 225
Otolith, *Anodonta* 183
Ovary, cockroach 90
crayfish 66
earthworm 15
frog 278, fig. 47, 279, 282
mussel 185
snail 226
Ovatella bidentata 239
Oxygastra curtisii 116
Oxyuris 92

Palæmon 60
Palmate newt 301
Paludestrina gen. et 5 sp. 236
Paludina vide sub *vivipara*
Paranais gen. et 1 sp. 34
Parasites, cockroach 92
of domestic animals 302–321
of earthworm 20
of wasps 155–159
Paste-worms 321
Pearls 164, 198–200
Pedal gland 225
Pegomyia inanis 147, 155, 156
Pelodera pellio 22
Periganglionic glands 271, 274
Periplaneta 71 et seq.
americana 95
Petrohyoid muscles 265
Physa, locomotion 214
fontinalis 238
Phytia myosotis 239
Pisidium definition of genus and 4 sp. 203
Plagiotoma blattarum 92
Planorbis 32; gen. et 11 sp. 238, 239
blood of 222
contortus 291
corneus 310
locomotion 214
Platychirus 79
Platycnemis pennipes 121
Platycleis brachyptera 97
grisea 97
Pollen brushes of bee 78
Polystomum integerrimum 290
Pomatias elegans 236
Pompilidæ 75
Porospora gigantea 69
Poultry, round-worms of 320, 321
tape-worms of 315
Pristina gen. et 2 sp. 36
Proglottis 311
Protozoa 302
Pseudamnicola anatina 236
Pulex serraticeps 313
Pulvillus 79
Punctum 242
Pupa vide sub *Jaminia*
Pupidæ vide sub *Vertiginidæ*
Pyralis farinalis 315
Pyramidula 242
Pyrrhosoma et 2 sp. 120

Rabbit 302, 313
tapeworms of 315, 316
Radula 216, 218, figs. 40, 41
Rana esculenta 292
temporaria 246
Rat 92
tapeworms of 315
Recuperative powers of earthworm 19

Red corpuscles 257
Redia 307
Regeneration, crayfish 68, 69
Reproduction, cockroach 89–91
 crayfish 65
 dragonfly 102
 earthworm 15–19
 frog 277 et seq.
 leech 30
 mussel 184–196
 snail 226–230
 wasps 150–152
Respiration, cockroach 83–85
 dragonfly 85, 105, 106
 earthworm 9
 frog 261–267
 hymenoptera 85
 leech 29
 mussel 178
 snail 220
Rhabdonema nigrovenosum 291
Rhipiphorus paradoxus 157–159
Ripistes gen. et 1 sp. 35
Round-worms 316–321

Sabre-tailed Grasshopper 96
Sand-wasps 123
Sawflies 123
Scales of Lepidoptera 87
Scaphognathite 56, 57
Sclerostomum equinum 320
Segmentina nitida 239
"Shamming dead" 86, 87
Sheep-rot 309, 310
Sinistral shells 206
Slavina gen. et 1 sp. 35, 36
Slugs 315
 hibernating 212
 shells of 212
Smell, sense of, cockroach 87
 sense in crayfish 65
Smooth Newt 300
Snail 205 et seq.
Snail-farms 232–234
Somatochlora et 2 sp. 116
Sounds produced by insects 88, 89
Sparganophilus gen. et 1 sp. 41
Spermatophore of earthworm 19
 snail 228
Sphæriidæ, definition of family 203
Sphærium, definition of genus and 4 sp. 203
Sphegidæ 123
Sphyradium 242
Spinal cord 271, 272, fig. 46
Spinal nerves 271
Spiral valve of leech 9
Spiral valve of dogfish 9
Spiroptera turdi 22
Sporocyst 306, 307
Stag-beetle, mandibles of 94
"Staggers" 313
Statocyst 60 et seq.
Statoliths 60 et seq.
Stenobothrus 95, 96
Stenogyridæ 242
Sternohyoid muscle 264
Steropus madidus 21
Stilesia 315
Sting of wasp 134–138
Strongylus 320
Stylaria gen. et 1 sp. 36
Stylodrilus gen. et 1 sp. 38, 39
Succinea gen. et 3 sp. 242
Suckers of tadpole 286, 288
Sword-tailed grasshopper 96
Symbius blattarum 92
Sympathetic System 274
Sympetrum et 5 sp. 114, 115
Syngamus trachealis 320

Tachydromia arrogans 79
Tadpole 285 et seq., fig. 49
Tænia cœnurus 313
 serrata 313
 solium 312, fig. 53, 313
Tapeworms 310–316
Testacella 21, 213, 216, 232
 gen. et 3 sp. 245
Tetronerythrin 58, 181
Tettix 95, 96
Thamnotrizon cinereus 97
Thread-worms 316–321
Thysanosoma 315
Toad 292
 breeding season 294
 skin 247
 spawn 283, 284, 294
Tracheal tubes 84, 85
Trematodes 70, 290, 291, 303–310
Trichina spiralis 317, fig. 54, 318

Trichodectes latus 313
Triton cristatus 300
Tubifex gen. et 3 sp. 37, 38
  hæmoglobin in blood of 10
Tubificidæ family 36
Tympanic membrane 264

Uncinaria duodenalis 318
  trigonocephala 319
Unio, genus, definition of 202
  margaritifer 199, 200, 202
  pictorum 202
  tumidus 203
Unionidæ, definition of family 202
Ureter, frog 275
Urinary bladder, frog 277

Vallonia 242
Valvatidæ fam. gen. et sp. 235
Vas deferens, frog 278, fig. 47
Vasa efferentia, frog 277
Vermiculus gen. et 1 sp. 37
Vertebræ 267–269
Vertiginidæ 243
Vertigo gen. et 8 sp. 243, 244
Vesicula seminalis, frog 279
Vespa austriaca 160, 161
  antenna-comb 77
  crabro 159
  diagram of nest 147
  germanica 160
  norwegica 161
  rufa 161
  sylvestris 161
  vulgaris 159, 160
Vespidæ 125
  list of British 159–161
Vinegar-eels 321
Vision of insects 88
Vitrea gen. et 8 sp. 240
Vitrina pellucida 239
Vivipara 222
  gen. et 2 sp. 235
Volucella bombylans 157
  inanis 156, 157
  zonaria 156

Wasps 123 et seq.
  cleansing apparatus 77
  food 141, 154
  genital armature 159
  homing power 152–154
  life-history 139–150
  mimics of 125
  movements of detached abdomen 85
  muscles for actuating wings 128
  nest 141–148
  parasites 155–159
  population of nest 148
  sex of larvæ 149–151
  sting 134–138
  wings of 126 (figure)
Wings, cockroach 72–75
  dragonfly 100, 110, 113, fig. 20
  of wasp, movements of 129, 130
Worm-castings, weight of 24

Xiphidium dorsale 97

'Yellow cells' of earthworm 11

Zonitodes 2 sp. 240
Zonitidæ 239
Zygopterides 113 (fig. of wings), 118–122

Printed in the United States
By Bookmasters